Value Management of Construction Projects

Value Management of Construction Projects

John Kelly
Steven Male
Drummond Graham

Blackwell
Science

© 2004 by Blackwell Science Ltd,
a Blackwell Publishing Company

Editorial Offices:
9600 Garsington Road, Oxford OX4 2DQ, UK
  Tel: +44 (0) 1865 776868
Blackwell Science, Inc., 350 Main Street, Malden, MA
02148-5020, USA
  Tel: +1 781 388 8250
Blackwell Science Asia Pty Ltd, 550 Swanston Street,
Carlton South, Victoria 3053, Australia
  Tel: +61 (0) 3 8359 1011

The right of the Author to be identified as the Author of
this Work has been asserted in accordance with the
Copyright, Designs and Patents Act 1988.

All rights reserved. No part of this publication may be
reproduced, stored in a retrieval system, or transmitted, in
any form or by any means, electronic, mechanical,
photocopying, recording or otherwise, except as
permitted by the UK Copyright, Designs and Patents Act
1988, without the prior permission of the publisher.

First published 2004 by Blackwell Science Ltd

Library of Congress
Cataloging-in-Publication Data
Kelly, John.
  Value management of construction projects/John
Kelly, Steven Male, Drummond Graham.
   p. cm.
  Includes bibliographical references and index.
  ISBN 0-632-05143-4 (alk. paper)
   1. Construction industry–Management.
  2. Value analysis (Cost control)
  3. Project management.   I. Male, Steven.
  II. Graham, Drummond.   III. Title.
  TH438.K435 2003
  690'.068–dc22
                                                    2003063026

ISBN 0-632-05143-4

A catalogue record for this title is available from the British
Library

Set in 10/12 Bembo
by DP Photosetting, Aylesbury, Bucks
Printed and bound in India by
Replika Press Pvt. Ltd., Kundli 131 028

For further information visit our website:
www.thatconstructionsite.com

Contents

Preface		*ix*
1	**Introduction**	**1**
	1.1 Definition	1
	1.2 Developments in value management	1
	1.3 Developments in UK construction	2
	1.4 Developments in value management practice	3
	1.5 The aims and objectives of this book	4
	1.6 References	7
Part 1	**Method and Practice**	**9**
2	**Developments in Value Thinking**	**11**
	2.1 Introduction	11
	2.2 Background	11
	2.3 The global development of value management	28
	2.4 Value engineering and value management: an overview of terminology and definitions	29
	2.5 The international benchmarking study of value management	36
	2.6 Conclusion	48
	2.7 References	48
3	**Function Analysis**	**51**
	3.1 Introduction	51
	3.2 Strategic function analysis: the mission of the project	51
	3.3 Strategies, programmes and projects	59
	3.4 Function diagramming	60
	3.5 Kaufman's FAST diagramming	66
	3.6 Functional space diagramming	70
	3.7 Elemental cost planning and elemental cost control	73
	3.8 Element function analysis	75
	3.9 Element function debated	77
	3.10 Conclusion	78
	3.11 References	80

4	**Teams, Team Dynamics and Facilitation**	**81**
	4.1 Introduction	81
	4.2 Groups	81
	4.3 Teams	83
	4.4 Team norms	86
	4.5 Team coherence	87
	4.6 Leadership	87
	4.7 Team development	88
	4.8 Team think	88
	4.9 Selecting team members	89
	4.10 Facilitation	90
	4.11 Facilitation defined	90
	4.12 Identity and role of the facilitator	91
	4.13 Facilitation styles	95
	4.14 Team composition	97
	4.15 Change management	98
	4.16 Conclusion	100
	4.17 References	100
5	**Current Study Styles and the Value Process**	**102**
	5.1 Introduction	102
	5.2 The value process	102
	5.3 Benchmarked study styles, processes and deliverables	105
	5.4 Other study styles	117
	5.5 Observations from practice	121
	5.6 Value studies: a revised process	140
	5.7 References	143
Part 2	**Frameworks of Value**	**145**
6	**Value Context**	**147**
	6.1 Introduction	147
	6.2 Defining value	147
	6.3 Value systems and clients to construction	153
	6.4 A strategic value management model	162
	6.5 The role of value management and value engineering	164
	6.6 References	167
7	**The Project Value Chain**	**169**
	7.1 Introduction	169
	7.2 Value managing projects	170
	7.3 The project value chain and the client value system	174
	7.4 The value thread within the single project value chain	177
	7.5 The decision to construct/decision to build	180
	7.6 Supply chain management in construction	184

	7.7 Creating value opportunities in the project value chain: value management (VM) and value engineering (VE)	190
	7.8 References	192
8	**Client Value Systems**	**195**
	8.1 Introduction	195
	8.2 Defining quality as part of value	195
	8.3 Quality systems	199
	8.4 Performance indicators	205
	8.5 A method for the discovery of the client's value system	206
	8.6 Conclusion	215
	8.7 References	216
Part 3	**The Future of Value Management**	**219**
9	**Professionalism and Ethics within Value Management**	**221**
	9.1 Introduction	221
	9.2 The value management knowledge base: founded on theory?	221
	9.3 The influence of the marketplace on the provision of value management services	234
	9.4 Value management: methodology, occupation or profession?	236
	9.5 Summary and conclusions	250
	9.6 References	255
10	**The Future of Value Management**	**257**
	10.1 The development of value management	257
	10.2 An enhanced VM process	258
	10.3 Value managing quality to deliver best value	263
	10.4 Value management futures	264
	10.5 Achievement of objectives in writing the book	272
	10.6 Reference	273
Appendix 1	**Toolbox**	**274**
	ACID test	274
	Action plan	275
	Audit	275
	Checklist	276
	Client value system	278
	Design to cost – BS EN 12973: 2000 *Value Management*	281
	Document analysis	281
	Driver analysis	281
	Element function analysis	282
	Facilities walkthrough	282
	Failure mode and effects analysis – BS EN 12973: 2000 *Value Management*	283

FAST diagramming (function analysis system technique) — 283
Functional performance specification – BS EN 12973: 2000 *Value Management* — 283
Functional space analysis — 283
Goal and systems modelling — 286
Idea reduction: judgement — 287
Impact mapping — 289
Interviews — 289
Issues analysis — 290
Kano — 291
Lever of value — 292
Life cycle costing — 293
Post occupancy evaluation — 293
Presentation — 294
Process flowcharting — 294
Project drivers — 294
Project execution plan (PEP) — 295
Quality function deployment — 296
REDReSS — 299
Risk analysis and management — 299
Site tour — 302
SMART methodology — 302
Spatial adjacency diagramming — 305
Strategies, programmes and projects — 306
Strengths, weaknesses, opportunities and threats (SWOT) — 306
Time, cost and quality — 306
Timeline — 308
User flow diagramming — 308
Value analysis – BS EN 12973: 2000 *Value Management* — 308
Weighting and scoring — 308
Whole life costing — 311
Appendix 1 references — 314

Appendix 2 **Case Study** — **315**
Introduction — 315
The case study — 315
Area community office project: brief — 316
VM study at elemental level — 333
An illustration of full value management — 337
Conclusion of case study — 362

Bibliography — 363
Index — 367

Preface

Value management is the name given to a process in which the functional benefits of a project are made explicit and appraised consistent with a value system determined by the client, customer or other stakeholders.

It is exactly ten years since the publication of our previous book *Value Management in Design and Construction: the Economic Management of Projects*, a period during which value management has developed into a defined service and has gained recognition within predominantly the manufacturing and construction sectors of industry and academia. In the UK there is an emergent but strong uptake of value management in the service sector led by the legislated demand for best value in public services.

The past ten years have largely seen the resolution of the debate over terminology aided by the publication of standards in many countries. It is generally accepted internationally that 'value management', or 'the value methodology' in the USA, describes the entire service. Other terms, principally 'value engineering', describe specific parts of the service. The year 2007 will mark 60 years of the life of the value management service. During the first four decades North American thinking dominated its development. However, during the past ten years developments in principally Europe, Australasia and China (notably Hong Kong) have seen the divergence of ideas and practices which have impacted the progress of the value methodology in North America. Interestingly, over the whole period value management has continuously improved, unlike other management fads that emerge, are applied with gusto and then die to be replaced by another.

Our past ten years have been spent researching, developing and applying value management. In 1993 John Kelly and Steven Male formed an informal joint venture with Drummond Graham of Thomson Bethune, Chartered Quantity Surveyors, Edinburgh, to develop and undertake value management consultancy. Since this time the team have been involved with over 200 value management studies for a variety of private and public sector clients for projects ranging from high profile national icon structures to public sector housing, undertaken at all project stages from inception to operation/facilities management and under all forms of procurement. We have also adapted and used the value methodology for partnering studies, introducing organisational change and assisting bid teams prepare for bid submissions. The latter has proved particularly useful on PFI, prime contracting and other new procurement routes requiring collaborative working and where large supply chain teams can often be brought together. We have also found the VM methodology particularly robust in undertaking project reviews and project briefing studies where we have assembled multi-disciplinary independent teams to work with us and alongside client representatives.

In 1996 the UK Engineering and Physical Sciences Research Council funded a team led by Steven Male and John Kelly to undertake an international benchmarking study of value management which resulted in a best practice framework document published in 1998. To date over 2000 copies of the benchmarked manual, accompanying research report and interactive CD-ROM have been sold internationally. In parallel, in-depth investigations of the underlying theories of value management as applied to construction projects have been the subject of a number of successfully completed PhD theses. This combined body of research work forms the foundation of the services, tools and techniques described in this book.

This book has been in development for some time. We have slowly but carefully recorded our own approach to the theoretical development of value management and subsequently tested each development through action research studies and consultancy.

As described in the Introduction, we have written the book to facilitate different modes of use. It will appeal equally, we trust, to the competent practitioner looking to benchmark an existing service; to construction clients, consultants and contractors looking to investigate value management further; and to the undergraduate and graduate student. Part 1 describes the background and structure of value management. The three chapters of Part 2 describe the theoretical framework within which value resides. Part 3 describes the service and its attendant professionalism and ethical issues, and also the likely futures. The extensive appendices contains a Toolbox of those techniques commonly associated with the practice of value management and also an illustrative case study.

During the past ten years we have interacted with several hundred construction clients, consultants, contractors and value management practitioners from many countries. We have taught undergraduate and postgraduate modules, run institutionally accredited value management courses and supervised research students at PhD level. In each activity we have gleaned, developed and tested theories, processes, procedures and practical tools. We are grateful to all the people and organisations too numerous to mention who have given us their time, views and insights. To any of them who think they recognise something here resulting from a comment or interaction with us we offer our special thanks.

John Kelly
Steven Male
Drummond Graham
July 2003

1 Introduction

1.1 Definition

Value management is the name given to a process in which the functional benefits of a project are made explicit and appraised consistent with a value system determined by the client.

This definition applies to all types of projects irrespective of which sector they come from. For example, the project could be the design and manufacture of a product, the design and construction of a building or infrastructure product, the re-evaluation of an organisational process or the provision of a new or improved service in banking, insurance or public services such as education or health. The factor that makes value management possible is the identification of the project. The client for the project will implicitly or explicitly establish a value system for the project. 'The client' in the context of this definition and for the remainder of this book is the person, persons or organisation responsible for the inception of the project and for its eventual adoption into the client's mainstream business.

This book is focused on construction; however, much of the thinking, philosophy, systems, tools and techniques are as applicable to projects in other sectors.

Maximum value as defined by Burt[1] is obtained from a required level of quality at least cost, the highest level of quality for a given cost or from an optimum compromise between the two. This is a useful definition because it highlights the relationship between value, quality and cost. In this book the definition of value is a relationship between time, cost and the variables that determine the quality the client seeks from the finished project.

1.2 Developments in value management

During its first four decades of life, value management developed within the manufacturing sector with only slight leakage into other areas. From the mid-1980s value management has been adopted for use as a value-for-money measure within construction industries of a number of countries. In the UK, the past two decades have seen growth in its development and practice at differing intervention points across a wide range construction project types. Over the same period risk management has developed and is often associated with value management as a complementary service. The developments in value management have reached a plateau and therefore this is an opportune time to encapsulate the developments in value thinking in construction and record robust models for the practice of value management in construction. That future developments in value management belong to the service sector is beyond dispute.

1.3 Developments in UK construction

A fertile ground was prepared for developments in value management in the UK construction industry during the 1990s by various initiatives that sought to increase the efficiency and effectiveness of the industry.

- The Latham Report[2] spawned the Construction Industry Board, which published influential works on the modernisation of the industry. Value management was seen to be conducive to good practice and received significant coverage.
- The Egan Report,[3] which spawned the Movement for Innovation (M4I), took advantage of web technology to showcase examples of good practice and provided an opportunity for benchmarking through its key performance indicator database. It was influential in shifting a substantial proportion of the construction industry towards more collaborative working, an environment in which value management thrives.
- The *Accelerating Change* Report[4] builds on the work of M4I and has established Rethinking Construction as the primary vehicle for public and private sector construction product and process advancement.
- The Office of Government Commerce has launched the Gateway process[5] which, with the accompanying construction procurement guidance, describes the benefits of good practice in construction procurement in the public sector. Documents describe the place for value management within this process and this is used for illustration later in this book.
- Newer procurement systems based on framework agreements, negotiations, guaranteed maximum price cost plus projects initiated forward-thinking contracts such as PPC 2000, Defence Estates Prime Contract and the NHS ProCure 21 procedure.

The climate of the 1990s in UK construction was therefore right for the development of innovative systems such as value management. The authors' research activity, which began in 1986 funded by the Education Trust of the Royal Institution of Chartered Surveyors (RICS), was boosted in the 1990s by further funding from the RICS and the Engineering and Physical Sciences Research Council. The latter funded a major study into the international benchmarking of value management practice that resulted in the completion in 1998 of *The Value Management Benchmark* published by Thomas Telford.[6]

The benchmark was the springboard for detailed work into two areas. First, to make clear study styles and their application at particular stages of projects and relate each study style with their most commonly associated tools and techniques. Second, to investigate the concept of quality and value to understand their interrelationship and their application within supply chain thinking. This research work was carried out by the authors or under their supervision using a variety of research methods. Significant findings were made through the grounded theory approach and action research.

1.4 Developments in value management practice

New practices, particularly in management, bring with them the bandwagon effect. A good idea is launched into the marketplace as a new service by an entrepreneurial consultant. Recognising the good idea, other consultants offer the same service. Over time, the service assumes the trappings of standardisation, regulation and institutionalisation; clients buy from the best, which can now be distinguished, and other consultants discontinue the service.

The history of construction project management in the UK can be tracked through landmark projects, such as the construction of the National Exhibition Centre in Birmingham. At this time employment of a consultant project manager was relatively rare but soon became recognised as a better way of doing business. Within a short period of time many consultancy organisations were selling the services of project management. However, over time those unique assets that make project management special have become recognised and the activity of project management has become a specific skill. The Association for Project Management now aims to be the first point of contact for and the national authority on project management, and through the International Project Management Association (IPMA), an international authority. It aims *inter alia* to develop professionalism in project management and to achieve recognised standards and certification for project managers.

Value management activity in UK construction began three decades after the early project management consultancies. Aided by the European SPRINT programme (strategic programme for innovation and technology transfer), a European standard for value management was published, authored by a consortium of the various value associations throughout Europe. A training and qualification system entitled Value for Europe has been configured with its own European Governing Board. Within the UK the Institute of Value Management is currently in the process of developing systems and procedures, ethics and standards and a branch network. The debate on professionalism is reserved for Chapter 10.

1.5 The aims and objectives of this book

In writing this book, the authors have brought together and synthesised the background, international developments, benchmarking and action research in value management to provide a comprehensive package of theory and practice. The book is overtly concerned with value management, examining function analysis and team dynamics and also proposing a method for determining the client's value system and quality criteria. The book examines different value management study styles and proposes solutions for various activities at different stages of projects. The book does not probe into the areas of creativity such as those described by De Bono or TRIZ or the fields of operational research and specifically operational hierarchies, nor does it address the whole subject area of group decision support. All these the authors leave to other academic colleagues.

The objectives of the book are to:

- Interpret the results of recent research and specifically the authors' own research into the international benchmarking of value management.
- Record accurately developments in value thinking during past decades, addressing the nature of value, transforming it into definitions and also discussing its alignment with total quality management and performance indicators.
- Examine the complexity of value systems that must be addressed in any VM study and specifically the project value chain and value thread.
- Present a reasoned argument for the development of the client's value system integrating the components of value, cost, time and quality.
- Present function analysis.
- Examine teams, team behaviours and facilitation and to point up practical issues when facilitating value management teams.
- Describe in sufficient detail for practical use a series of VM study styles, tools and techniques.
- Describe an enhanced VM process, argued to be the potential foundation for future 'professional' development.

The authors intend that this book facilitates the gaining of value management knowledge in a number of ways:

- For dipping in and looking for a particular topic use the Contents list at the front of the book or the index.
- For checking on a particular value management tool or technique – the Toolbox (Appendix 1) is provided in alphabetical order.
- For an understanding of the process of value management, Part 1, together with the Appendices, will give sufficient information for the background and development of value management to be appreciated together with the study styles, tools and techniques which combine to form a value management service.

- For an accomplished value management practitioner to benchmark their service, Part 1, Chapter 5, together with the Appendices give the study styles, tools and techniques, described to permit the practitioner to adopt or amend them for their own personal use. Additionally, Part 2 explores the value concepts and describes a method for the construction of the client's value system.
- For a theoretical overview of value the reader is referred to Part 2, in which the authors expose value and break it down into a number of discrete points and themes.
- For the authors' thoughts on the subject of professionalism and ethics, see Part 3.
- The conclusion gives the opportunity for both pulling the book together and 'flying some kites' regarding the future of value management. In this section the authors have prepared three scenarios and have indicated the future of value management within each.

The book is written for a number of audiences: for the competent practitioner who may be looking to benchmark their existing service; for construction clients, consultants and contractors who may be looking to probe value management further; and for the undergraduate or graduate students.

For the undergraduate or graduate student taking a course or module that includes value management this book is designed both to be an exposition of the process and to present some fertile ground for individual thought.

For the researcher, this is the authors' view of the value story thus far. There is still more work to do, particularly in understanding the adding of value to products and services through the value chain as well as the retaining and nurturing of the value thread that starts with the client and weaves its way through the fabric of the project. We have utilised numerous research methods since the mid-1990s including hypothetico deductive, action research and grounded theory based analysis. Techniques have included literature analysis, case study analysis, case vignettes, benchmarking, questionnaire survey, structured and unstructured interviews and the real-time analysis of live projects.

The early years of value management have been dominated by US practice. A watershed occurred in the mid-1980s with the international use of the method in construction. Whilst there has been interest by some countries in taking forward VM by franchising the US methodology, it was also taken and melded into a diverse range of international construction markets and cultures. The chapters in this book weave a rich a tapestry of insights into value thinking drawing threads from North American practice, the International Benchmarking Study of Value Management and a programme of action research undertaken by the authors.

Chapter 2 provides an overview of American value practice and a detailed review of North American study styles, and subsequently introduces insights from Japan, Europe and Australia.

Chapter 3 describes approaches to strategic function analysis together with a functional approach to elements and components, drawing the distinction between function analysis methods applicable to value management and those applicable to value engineering.

Chapter 4 summarises current thinking in respect of groups and teams, and outlines the characteristics and methods for the facilitation of value management and value engineering teams. The chapter draws a distinction between groups and teams, reflecting that teams, particularly for value studies, are likely to be a *fait accompli* rather than a group of like-minded people or a team crafted to comprise a given set of individual attributes.

Chapter 5 presents a synthesis of Chapters 2, 3 and 4 as an overview of the value process including developments resulting from recent research. Different value study styles are introduced pursuing the argument that the role of the value manager is one of deciding on, structuring and delivering a study style tailored to a particular value problem, be it for a project, project programme, service or organisational function. Irrespective of the type of value problem, it is postulated that the stages in its solution comprise three generic phases:

- The orientation and diagnostic phase
- The workshop phase
- The implementation phase

Chapter 6 discusses value and value systems, defining 'value management' and 'value engineering' as used by the authors. The chapter argues that value management is business project or investment focused, whilst value engineering, as a subset of value management, is more technical project focused. The chapter sets out a typology of clients to introduce the idea of different client value systems, each of which must be understood in order for a value study to be tailored, and concludes by raising some ethical issues.

Chapter 7 describes the value theory that provides the raw material for the project value chain, drawing on an action research programme of studies that has been underway since the early 1990s. It describes the application of value chain theory to value management and value engineering projects throughout the project life cycle; procurement studies that have encompassed the Private Finance Initiative (PFI), prime contracting, partnering and other procurement routes; and, finally, organisational development projects that have involved setting up team structures, reviewing departmental organisational structures and introducing change into organisations. The chapter also integrates recent work undertaken by Male[7] on supply chain management.

Chapter 8 presents a synthesis of Chapters 6 and 7, drawing together the theories of value, value systems and the project value chain and melding this with the theories of quality, based upon the accepted concept that value is a function of cost, time and quality. Work in total quality management is reviewed along with quality based performance indicators to evolve an operational technique for the discovery of the client's value system. Three operational techniques are discussed.

Chapter 9 focuses on the service of value management considering the ethics and professionalism of the value manager in practice. It explores some of the developmental issues now facing value management as a methodology. Building on the earlier chapters it looks at the market settings within which practice takes place, explores occupations, professions and different models of professions, and assesses the implications of this analysis for value management.

Chapter 10 concludes the book and looks, through three scenarios, at the future for value management.

The Appendices contain a value management Toolbox – an alphabetical listing of those tools and techniques commonly associated with value management – together with a case study illustrating the use of the techniques. Further, while there are a number of excellent books on the subject of whole life costing, it was decided in this book to include a short overview of the principles and the principal formulae involved in life cycle costing as a reference since value management is increasingly practised in an environment of whole life appreciation.

The chapters in this book demonstrate the contextualisation of value management within construction and are principally influenced by a UK-style construction culture. The book draws together developments in value management thinking and practice. Value management is still developing: the aim of this book is to draw a line in the sand to denote where we have got to. There's still a way to go.

1.6 References

1. Burt, M. E. (1975) *A Survey of Quality and Value in Building*. Building Research Establishment, Garston, Watford, Herts.
2. Latham, M. (1994) *Constructing the Team*. The Stationery Office, London.
3. Egan, Sir John. (1998) *The Egan Report: Rethinking Construction*. Department of the Environment, Transport and the Regions, London, Publications Sale Centre, Rotherham, England.
4. Strategic Forum for Construction (2002) *Accelerating Change*. Rethinking Construction, London.
5. Office of Government Commerce (2003) *Gateway to Success*. The Stationery Office, London.
6. Male, S., Kelly, J., Fernie, S., Gronqvist, M. & Bowles, G. (1998) *The Value Management Benchmark: A Good Practice Framework for Clients and Practitioners*. Published report for the EPSRC IMI Contract. Thomas Telford, London.
7. Male, S. P. (2002) Supply chain management. In: *Engineering Project Management* (ed. N. Smith), 3rd edn. Blackwell Publishing, Oxford.

Part 1 Method and Practice

The value methodology as a structured management service has been recognised for over 50 years. It began in the manufacturing industry of the USA in 1947 and migrated to the construction industry in the late 1960s. In the UK, value management in construction evolved in the late 1980s. The first UK textbook *Value Management in Design and Construction*[1] by the authors was published in 1993. Since that time value management in construction has evolved to become an established service with commonly understood tools, techniques and styles. The authors undertook a UK Engineering and Physical Sciences Research Council funded research project entitled 'the International Benchmarking of Value Management' between 1996 and 1998 to understand, from an international perspective, the tools, techniques and styles. From this data and from ongoing action research in value management the authors have concluded that value management is a service with three primary core elements, a value system and a team-based process, with function analysis promoting understanding. It is a process with three generic phases:

- The orientation and diagnostic phase.
- The workshop phase.
- The implementation phase.

It is the authors' view that the role of the value manager is to structure a study strategically and tactically to take account of these phases within it. The skills of the value manager comprise the ability to understand a value problem, structure a process to bring value systems together and introduce improvements. Value management is, therefore, a change-orientated process and needs to be treated, designed and delivered as such.

Part 1 comprises four chapters giving the background and structure to the practice of value management. Chapter 2 addresses the developments in value management and value engineering, reviewing practice in the USA and contrasting it with that in Japan, outlining guidance material from the UK and the Australian public sectors and giving an overview of European/British Standard BS EN 12973:2000 on value management. Chapter 3 addresses function analysis, considered fundamental to the practice of value management and value engineering. Chapter 4 is a review of the literature on teams, team dynamics and facilitation, weaving the theoretical threads into a working model for use in a value management context. In Chapter 5 the authors discuss the value process and study styles distilled from benchmarking and action research and consolidated into frameworks for practical application. Lessons learned from practice are highlighted, together with many factors of which the practising value manager should be aware. These four chapters give the basics of value management. Part 2 addresses the value context, the project value chain and the client value system. Appendix 1 provides a Toolbox of the techniques referred to in the text.

Reference

1. Kelly, J. & Male, S. (1993) *Value Management in Design and Construction*, E. & F. N. Spon, London.

2 Developments in Value Thinking

2.1 Introduction

The authors would contend that developments in value thinking from 1947 to the late 1980s have been dominated by American thinking. This chapter provides an overview of American value practice, a detailed review of North American study styles and subsequently introduces insights from Japan, Europe and Australia. Other chapters in the book weave a rich tapestry of insights into value thinking, drawing threads from North American practice, the International Benchmarking Study of Value Management and a programme of action research. Chapter 5 presents current study styles and the value process.

2.2 Background

Value management, a term which was never widely used in the USA, came to signify a service in which more than just the basic function of an element or component was to be considered. Value management is the structured management of the total value equation through all stages of the project and therefore in this respect subsumes value engineering as a component part of the whole service. The UK Value Engineering Association, established in 1966, changed its name to the Institute of Value Management in 1972. The more recently formed Institute of Value Management Australia and the Hong Kong Institute of Value Management also see value management as an activity that is wider than value engineering. SAVE International in the USA has opted to use value methodology as the all-inclusive term.

Value thinking began with the work of Lawrence Miles who, in the 1940s, was a purchase engineer with the General Electric Company (GEC). At that time manufacturing industry in the United States was running at maximum capacity, which resulted in shortages of a number of key raw materials and components. GEC wished to expand its production and Miles was assigned the task of purchasing the materials to permit this. Often he was unable to obtain the specific material or component specified by the designer so he reasoned, 'If I cannot obtain the specified product I must obtain an alternative which performs the same function.' A characteristic from the beginning was the team approach to creatively providing the required function, resulting in the generation of many alternatives to the existing solution. Where alternatives were found they were tested and approved by the product designer.

Miles found that many of the substitutes were providing equal or better performance at a lower cost and based on these observations he proposed a system that he called value analysis. The definition of value analysis is:

> An organised approach to providing the necessary functions at the lowest cost.

Value analysis was seen to be a cost validation exercise that did not affect the quality of the product. The straight omission of an enhancement or finish would not be considered value analysis. This led to the second definition:

> Value analysis is an organised approach to the identification and elimination of unnecessary cost.

Unnecessary cost is:

> Cost which provides neither use nor life nor quality nor appearance nor customer features.

In this context:

- Use refers to the utility of the component. This utility is measured by reference to the extent to which it fulfils the required function.
- The life of the component or material must be in balance with the life of the assembly into which it is incorporated. For example, unnecessary cost may be incorporated if a component is specified which has a life of twelve years within a product that will be redundant in four years.
- Quality is subjective. However it is perceived, it must be preserved. The philosophy of value engineering looks towards reducing cost without sacrificing quality. Quality is discussed further in a later chapter.
- The appearance of a product is often one of the most important features to a customer. The aesthetic features of any building or product are often those that delight and attract the customer.
- Customer features are those that sell the product but which of themselves do not add to the functions. Graphic or pictorial designs applied to products are a good example.

In 1954 the US Department of Defense's Bureau of Ships became the first US Government organisation to implement a formal programme of value analysis. It was at this time that the term value engineering came into being for the administrative reason that engineers were considered to be the personnel most appropriate for this programme. The formation of the Society of American Value Engineers in 1959 established value engineering, which came into common use as the preferred term and is the term most used in the USA today. The Society of American Value Engineers became SAVE International in 1977. The original term of value analysis is still widely used in France.

US Government patronage

Important initially to the development of value engineering in the USA was the involvement of government that necessarily demanded auditable structures. Value engineering spread to many US federal, state and local government agencies following the introduction of the cost reduction programme of Secretary McNamara in 1964. Within construction, government agency action was instrumental in the furtherance of the technique. Tender documents issued by, *inter alia*, the US Department of Defense, Navy Facilities Engineering Command, required value engineering to be carried out at a specific stage in the development of a project, according to a set 40-hour programme called the Job Plan using a team independent of the project team.

A summary of US Government patronage is shown in Table 2.1

Table 2.1 Summary of US Government agency patronage.

Year	Description
1963	Introduced into US Department of Defense, Navy Facilities Engineering Command.
1965	Introduced into US Department of Defense, Army Corps of Engineers.
1968	Used by Facilities Division of the National Aeronautics and Space Administration.
1973	US General Services Administration, Public Building Service publish the first VE service contract clauses for use in Design and Construction Management contracts.
1974	US General Accounting Office publish (May 6, B-163762) 'Need for increased use of value engineering, a proven cost saving technique, in federal construction'.
1976	US Environmental Protection Agency, Water Program Operations, publish a 'VE workbook for construction grant projects' MCD-29, making VE mandatory on projects over $10 million.
1993	US Office of Management and Budget (OMB) issue a circular calling for 'government-wide use of value engineering and requiring Federal agencies to implement value engineering techniques'.
1996	President Clinton signs the Defense Authorisation Act, now known as public law 104–106, which states in Section 36 that each executive agency shall establish and maintain cost-effective value engineering procedures and processes. It further states that in this context the term 'value engineering' means an analysis of the functions of a program, project, product, item of equipment, building, facility, service, or supply of an executive agency or contractor personnel, directed at improving performance, reliability, quality, safety and life cycle costs.

14 Method and Practice

The job plan

A characteristic of North American value engineering is the team approach to function definition and creativity through application of the job plan, a logical, sequential, approach to the study of value. The early job plans were manufacturing orientated as the example below but the principles of the job plan and its name, for better or worse, have been applied in other sectors.

All North American value engineering texts refer to a pattern derived from Miles's original job plan which is summarised as follows[1]:

Phase 1: orientation. Asking: what is to be accomplished, what does the client need and/or want, what are the desirable characteristics?
Phase 2: information. Secure all costs, quantities, drawings, specifications, manufacturing methods, samples and prototypes. Understand the manufacturing process. Determine the amount of effort that should reasonably be expended on the study.
Phase 3: speculation. Generate every possible solution to the identified problem using brainstorming sessions. Record all suggestions.
Phase 4: analysis. Estimate the dollar value of each idea and rank in order of highest gain and highest likely acceptability. Investigate thoroughly the best ideas.
Phase 5: programme planning. Establish the manufacturing programme by identifying operations, design and production personnel, suppliers, etc. Promote an ethos of creativity in all involved parties.
Phase 6: programme execution. Pursue the programme, evaluating and appraising further suggestions from suppliers, etc.
Phase 7: status summary and conclusion. If in a position to take executive decisions, act on new ideas; if not, make recommendations to those who are to make the decisions.

It is to be noted that Miles was solely concerned with manufacturing and this is reflected in his job plan. Those subsequent authors whose concerns included construction modified the job plan to accord with the processes and terminology of the industry. The review which follows contains the essential ingredients of the North American job plan applied to construction.

Phase 1: Orientation

The orientation meeting, promoted by the General Services Administration and practised by New York City Office of Management and Budget[2,3], is held following the appointment of the design team. It is a meeting chaired by the value engineer for the project and attended by the design team and those client representatives who have an interest in or possess some ownership of the problem being addressed. The objective of the meeting is to pose the questions asked by Miles, namely:

- What is to be accomplished?
- What does the client need and/or want?
- What are the desirable characteristics?

As practised by New York City, it is an opportunity to allow everyone involved in the project to understand all the issues and constraints. It provides everyone who is to make a decision with an opportunity to give and receive information.

Phase 2: information

In this phase all of the available information relating to the project under review is gathered together. The objective of the information gathering is to identify the functions of the whole or parts of the project, as seen by the client organisation, in clear unambiguous terms. The information should not be based upon assumption but be obtained from the best possible source and corroborated if possible, with tangible evidence. The reasoning behind this is that the quality of decision making cannot rise above the quality of the information upon which the decision is to be made.

The specific information being sought at this stage is:

- *Client needs*, which are the fundamental requirements that the project must possess to serve the client's basic intentions. Needs should not be seen solely in terms of utility as the client may have a need for a flamboyant statement or a need for a facility which heightens the client's esteem.
- *Client wants* are the embellishments which it would be nice to have but do not satisfy need.
- *Project constraints* are those factors that will impose a discipline upon the design, for example, the shape of the site, planning requirements, regulations, etc.
- *Budgetary limits* expressed as the total amount that may be committed to the project in initial capital and life cycle costing terms.
- *Time* for design and construction as well as the anticipated period for which the client will have an interest in the building.

Although, as stated above, the quality of decision making cannot rise above the quality of the information on the basis of which the decision is to be made, care should be taken not to spend unjustifiable time and effort in information seeking. Fallon[4] refers to the dilemma between the dangerous consequences of acting upon inadequate information and the possible missed opportunity while waiting for reliable information to arrive.

As an aid to information gathering, some authors consider that the value engineering team should construct a function analysis system technique (FAST) diagram. This function logic diagram, resembling a decision tree, is commenced with the prime function to the left and constructed to answer the questions 'why?' when reading from right to left and 'how?' when reading from left to right. It is argued that the very action of drawing the FAST diagram concentrates the minds of the value engineering team on the functional requirement of the project.

The concept of function and the FAST diagram is considered in Chapter 3.

Whether or not a FAST diagram is used, a prime task in the information stage is the realisation of those items in the current brief or design which attract high cost for low functional utility and/or those items which have a high importance but attract a low cost.

North American practice has evolved a tradition that functions be recognised in a simple verb plus noun form. A wall, for example, may have the following functions: support roof; exclude weather; maintain internal environment. It is these functions that the value engineering team will answer in terms of technical solutions.

Phase 3: creativity

In the creative phase the value engineering team put forward suggestions to answer the functions that have been selected for study. Normally only a few cost dominant functions will be selected for study.

There are a number of creative techniques, for example brainstorming, the Gordon technique, the Synectics technique and many more.

Phase 4: evaluation

The value engineering team evaluate the ideas generated in the creativity phase using one of a number of techniques, many of which depend upon some form of weighted vote. This stage forms a crude filter for reducing the ideas generated to a manageable number for further study.

Phase 5: development

The accepted ideas, selected during phase 4, are investigated in considerable detail for their technical feasibility and economic viability. Outline designs will be worked out and costs realised. There is wide scope for the use of life cycle cost models and computer aided calculations at this stage.

At the end of the development stage the team will again consider the worked-up ideas. Those that do not comply with the basic value engineering philosophy will be dismissed; that is, all ideas which either cost more than the original or are found to reduce quality are rejected.

Phase 6: presentation

The refined ideas supported by drawings, calculations and costs are presented by the value engineering team to the body that commissioned the value engineering exercise.

Phase 7: feedback

It is important that the value engineer receives some detail of those ideas that have been put into practice and is given the opportunity of testing the design and cost predictions of the team.

All North American texts on the subject of value engineering refer to, develop and discuss the job plan. Its principles are judged to be sound; indeed it resembles in many respects the classical *modus operandi* of research. The job plan is considered to be the foundation of a recommended value management methodology and therefore will be frequently revisited.

Function, value, cost and worth

These four terms are defined within the context of North American value engineering as follows:

- *Function* is a characteristic activity or action for which a thing is specifically fitted or used or for which something exists. Therefore something can be termed functional when it is designed primarily in accordance with the requirements of use rather than primarily in accordance with fashion, taste or even rules or regulations. Value engineers distinguish between a basic function and a secondary function. A basic function is defined as the performance characteristics that must be attained by the technical solution chosen. Secondary functions are the performance characteristics of the technical solution chosen other than the required basic function. For example, a basic function of a window may be to transmit light. A technical solution to the function could be a sheet of glass. Secondary functions of the design required because of the choice of a sheet of glass as the solution are to prevent solar glare, to control heat gain, to prevent condensation and to prevent cold radiation.
- *Cost* is the price paid or to be paid. It is often said that one man's price is another man's cost.
- *Value* is a measure expressed in currency, effort or exchange or on a comparative scale which reflects the desire to obtain or retain an item, service or ideal. In many texts the relationship of value to function and cost is represented by the expression:

$$\text{value} = \frac{\text{function}}{\text{cost}}$$

The concept of value is discussed in detail in Part 2.

- *Worth* is defined by North American value engineers as the least cost to perform the required function or the solution that will deliver the required function at the least cost.

The definitions above are a summary of the views of authors of North American value engineering texts in that none hold an opposing view. However, there is considerable room for discussion particularly relating to the extent to which functions can be priced. It may be the case that functions can be appreciated in terms of cost but not actually costed. For example, a window that transmits light through the medium of a sheet of glass may be costed. The additional cooling load required to control heat gain through the glass may also be costed. The alleviation of the cold radiation feeling of someone whose desk is next to the window is less easy to cost. The psychological benefit of being able to see outside or the aesthetic contribution to the structure as a whole is impossible to cost. Equally, is 'worth' as defined above an appropriate definition?

North American study styles

Kelly & Male[5] identified four formal approaches to North American value engineering defined as:

(1) *The charette* is a meeting following the compilation of the client's brief, attended by the full design team and by those members of the client's organisation who have contributed to the brief. This meeting is conducted under the chairmanship of the value engineer.
(2) *The 40-hour study* is an examination of the design developed to sketch design stage. This is carried out by an independent team of design professionals who have not been involved with the design until the time of the study, again under the chairmanship of the value engineer.
(3) *The value engineering audit* is a study of the proposals made by a subsidiary of a large holding company for a vote of capital to fund a project. This study will be undertaken by a value engineering team in order to ensure that the parent company is receiving value for money.
(4) *The contractor's change proposal* arises when a clause in the construction contract allows the contractor to suggest changes to the proposed design in order to reduce construction costs. The contractor receives a bonus in exchange for the proposal.

While the data upon which these four approaches were identified dates from 1993, recent discussions with North American value practitioners have confirmed that these styles are still representative of value engineering. Tension has been observed in the context of the 40-hour workshop, which is seen by some practitioners as being the mainstay of value engineering practice and forms the principle framework for delivering training. The tension arises from the cost of the study generally offered to the private sector. In order to reduce costs the shortened version of the 40-hour study reviewed in 1993 is becoming more common and a number of practitioners are also engaging with the design team –

'the design team of record' – rather than employing a completely independent team. The 40-hour study procured by the public sector provides a competitive basis for a standard service, involving an independent team that also provides an element of public accountability.

Comparing UK and US project procurement practice

In traditionally procured construction in the UK, the architect is the design team leader with responsibility for taking the brief and the budget from the client, preparing sketched alternatives, and discussing the financial consequences with the chartered surveyor and engineering matters with the various engineering consultants. The chartered surveyor is responsible for preparing the cost plan which should equate with the budget, financially controlling the evolving design by calculating the cost of the design as it is developed by the architect and the various consultants, and comparing this with the cost plan. Any dangers of overspend result in reference back to the design consultants for an alternative design solution. The chartered surveyor also prepares the tender documentation that will contain a detailed description of the works but often not a full set of working drawings.

In North America there is no chartered surveyor. The engineer enjoys a much higher status than in the UK and may be the design team leader in place of the architect. The architect/engineer has responsibility for taking the brief and the budget from the client and is responsible for preparing a cost estimate. Often this cost estimate is sub-contracted to a cost consultancy, employed by the architect/engineer to estimate the tender value based upon the sketch drawings. There is no financial control service. The cost consultancy has very little status and employs in the main personnel who have trained as contractor's estimators. This trend is one that is slowly changing and the term quantity surveyor is not unknown in North American construction.

In North America, during the development of the sketch scheme the architect and engineers work in parallel, so that at sketch design the architect's and engineers' draft schemes are available. This is in contrast to the UK system, where engineering installations are commonly designed after the architect's scheme is complete.

North American tender documentation comprises drawings and specifications often including performance specifications for work to be designed by the contractor. The North American system tends to be very linear with little scope for change, in contrast to the UK-style approach that iterates and has the appearance of being designed for change.

The position in the UK, however, has changed since the mid-1990s with the introduction of private finance into public projects through BOOT (build own operate transfer) schemes such as the Private Finance Initiative (PFI) and other similar initiatives for framework agreements and partnering. In the private sector, the design and build form of procurement, with or without framework agreements, and partnering have become more common. These changes have redistributed the time and cost risks from the client to the service provider, commonly the contractor.

The formal approaches to North American value engineering

Approach 1: The charette

This method seeks to rationalise the client's brief through the identification of the function of key elements and the spaces specified. This analysis through function at a meeting involving the client's staff and the design team should ensure that the latter understand fully the requirements of the former. During the research into this subject interviews were held with a client who had commissioned a value engineering study following a tender in which the lowest was 18% higher than the permitted budget. The client stated that a value engineering study at briefing would be useful

> '...just to focus on what are the objectives, what are the primary functions of any given activity. And this is one area of value engineering where it was clear when we got around the table after the design was done, including the players who were involved in the design, that what each of them thought was the primary function was different – and this was after the design was done.'

It should be noted that this particular study revisited the brief following the receipt of the tender. Although grossly inefficient in terms of the time spent on redesign it did enable the project to become viable. A true charette will study the brief following its compilation.

The brief given by the client to the design team is generally an amalgam of the 'wish lists' of all of the parties who contribute to the brief. This is particularly so for buildings which are to house organisations comprised of diverse departments such as hospitals, universities, prisons, owner-occupied offices for complex organisations and manufacturing plants. Even where a project manager is employed there is a high probability that the brief will reflect data gathered from departmental heads who will seek to maximise their requirements. In a prison, for instance, two departments may each have a requirement for a therapy area and the two areas may be identical, but this is not likely to be realised unless a study is made of the function of each space.

The charette is organised along the traditional job plan lines with the first stage being to gather as much information as is available regarding the function of the spaces defined in the brief. The functions of all of the spaces are defined along with performance criteria, e.g. this space must be held at a constant $20°C \pm 2°$ where the activity within the space generates heat.

The next stage in the process is to be creative in terms of arrangement, adjacency, timetabling, etc. It may be found, for example, that by siting two particular hospital departments together they may use the same laboratory with immense savings in capital and running costs (including laboratory staff).

The ideas generated are recorded and analysed and the final decisions are incorporated into the brief.

The charette has five major advantages:

- First, it is considered by many clients to be inexpensive. There is of course staff time to consider and the time of the design team, but the only real expenditure is the fee of the value engineer himself. One client interviewed during this research also stated that it was beneficial to have a cost consultant at this meeting. There is finally the secretary to the team who will be provided either by the client or by the value engineer.
- Second, the exercise was considered by some clients to be the best way of briefing the whole team. One industrial client with a large and expanding building stock stated that, even if the exercise did not realise any great rationalisation, the very fact that all members of the team, the architect, structural engineer, mechanical and electrical engineers, etc. were present meant that all fully understood his requirements.
- Third, the exercise occurs early in the process, stated by many to be the most cost consequent stage.
- Fourth, the exercise can be carried out in a short period of time; only the most complex projects will involve more than two days' work.
- Finally, the exercise cuts across organisational, political and professional boundaries. One central government organisation client stated that a meeting of this kind would not normally be possible since departmental heads would be reluctant to give up the time. Also the meeting, if held, would be politically structured. The design team themselves would not normally be sufficiently proactive in the organisation of such a meeting. The fact that it was a value engineering exercise under the chairmanship of an outsider made it happen.

The charette is therefore an inexpensive means of examining the client's requirements by the use of function analysis and allowing rationalisation and full design team briefing. The charette is one of the study styles that has carried through to UK practice and is described in Chapter 5.

Approach 2: The 40-hour value engineering study

The 40-hour study is the most widely accepted formal approach to value engineering; indeed the initial training of value engineers as laid down by SAVE International is based on a 40-hour training workshop. The study involves the review of the sketch design of a project by a second design team under the chairmanship of a value engineer. It applies all of the stages of the job plan within a working week and is seen by those operating it in USA as being quick and effective. The 40-hour study is described in detail here as being the form most often used, certainly in the US public sector. The procedure for the study is as follows:

- The client should inform the members of the design team at the time of their fee bid that the project will be the subject of an independent value engineering exercise. This is important both from a human relations aspect and also from the point of view of establishing how the design team are to cover the cost of any redesign work arising out of the exercise. Some clients require the members to cover this cost within their fee bid. Others state at the time of the fee bid that the design team members will be reimbursed for any necessary redesign work on an hourly basis.
- The client appoints the value engineer (VE), and in discussion with the design team establishes the date for the study. Normally the VE will submit a fee bid that covers the cost of the complete value engineering exercise described below.
- The VE will appoint a value engineering team, normally six to eight professionals in a mix that reflects the characteristics of the project under review, so for instance a project with a large amount of mechanical and electrical servicing may attract a team including four members with these professional backgrounds. These team members will be drawn from professional practice and may or may not have any previous value engineering experience. The team members are paid by the VE.
- The study is normally held near the site of the proposed project, either in a hotel or in a room provided by the client within the client's office.
- The date of the study is a key date for the design team and the value engineering team. The design must be complete to sketch design stage one week before the date of the study. This includes the architectural design and also the structural, mechanical and electrical engineering designs. The completed drawings are sent to the VE for distribution to the team during the week preceding the study.

During the week of the study the team will follow strictly the stages of the job plan. It is the logical step by step approach to the generation of alternative technical solutions which makes value engineering unique.

Monday: phase 1 – information
The team have each had the project sketch drawings, initial cost estimate together with calculations and outline proposals for the structure and services for two days and will have gleaned some information from these. At the beginning of the study the architect and the client are present. The VE gives an introduction and states the objectives for the week. Often the VE will have prepared a timetable and may also have prepared an elemental breakdown of the initial estimate.

Following the introduction the client and the design architect present the project and answer questions, and the client reaffirms which areas of the project are within the scope of the exercise. This latter point is important since if, for example, the client has already reached an agreement with a trade union that a specific number of people will be employed within a plant then all ingenuity on the part of the value engineering team to reduce manning levels will be in vain.

After the presentation the client and architect leave.

The team now concentrate their efforts on identifying the functions of the various parts of the building. In the study emphasis is given to those functions that are not important, or are secondary, but attract a high cost. Attention will also be paid to those functions that are primary and important but attract a low cost.

In one study for the modernisation of a boiler house on a large military site in North America, with an estimated project cost of $71 500 948, the team identified 17 functions of which 7 were selected for study.

Tuesday afternoon: phase 2 – creativity
During this phase the group brainstorm ideas to satisfy the identified functions. In the boiler house example above over 200 ideas were generated during this session. Creativity is a rapid but exhausting process: 200 ideas could easily be produced in one hour.

Tuesday afternoon: phase 3 – judgement
At this stage the team decide which of the ideas generated are worthy of further development. For example, of the 200 plus ideas generated above only 42 were thought good enough for development.

Before moving on to the development phase some value engineers prefer to invite the design architect back to the study at this point to discuss the acceptability of the ideas in principle. This can reduce abortive work if, for instance, the design team had already thought of the idea and rejected it or if the architect would not agree to such an idea under any circumstances.

Wednesday and Thursday: phase 4 – development
During this phase the team may split into individual or small groups to work on the ideas in detail. The aim is to develop the ideas into technically viable and costed solutions.

Friday: phase 5 – recommendation
In the morning of the final day the group reconvene to discuss the worked solutions. At this stage those solutions that either cost more than the original, reduce quality or are not technically feasible are rejected. In the boiler house case above 15 worked solutions were rejected at this stage leaving 27 viable solutions for presentation to the client and design architect.

In the afternoon those worked solutions accepted by the team are presented to the design architect and the client.

The formal study is now at an end. The members of the study return to their practice leaving the VE to take away the week's work and write the report.

The following week: action and feedback

In the early part of the following week the completed report is sent by the VE to the client and the design architect. At this stage, one North American government department takes all of the ideas and sets them out on a sheet horizontally with vertical columns for each member of the design team who receives a copy. The team members are requested to enter either 'accepted', 'rejected', or 'further discussion required' against the suggestions. A meeting is called where all members of the design team gather to discuss the suggestions. All those that have been unanimously accepted are required to be incorporated into the design. In respect of the others discussion takes place to determine which may be acceptable. The client will wish to be convinced of the need for rejection.

In the boiler house example above, 11 of the 27 suggestions were incorporated into the final design, leading to $32 868 302 savings on the original estimate of $71 500 948 with no loss of quality of function. This remarkable saving of 45% of estimated cost was achieved by demolishing two perfectly satisfactory buildings adjacent to the site and rebuilding them elsewhere. Within the original scheme the design team had used considerable ingenuity to design an expansion in the boiler house facility around the constraint of the existing buildings.

Advantages and disbenefits of the value engineering study

A value engineering study is seen as being effective by reason of:

- The generation of alternative technical solutions to a problem that has been costed in initial and life cycle cost terms.
- The fixing of a date for the completion of a sketch design. Although not a function of the study, it has been stated that the setting of a date for the study forces the design team to complete to a more advanced stage than would otherwise be achieved particularly by the engineering designers.
- In the majority of cases the costs of the study are a small proportion of the savings achieved. Value engineers state that on any project at least 10% of the estimated contract value is within the area of unnecessary cost. They also state that the value engineer will achieve a 10:1 return on the investment made by the client and therefore in the majority of cases the study fee, usually quoted as a lump sum, would work out to be not more than 10% of the savings realised.

In responding to questions on 'how often do studies fail?', value engineers stated nil but one major client stated about 2%.

The problems associated with the study relate to conflict, time and resourcing.

- The client may consider that the design team should arrive at the optimal solution without the need for a further exercise at additional expense. This criticism may be countered in two ways: first, that it is the function of the design team to arrive at a workable solution given the information in the client's requirements; secondly, that a value engineering study is an analysis of the ideas which have been generated. A value engineering study cannot be carried out until there is an idea to analyse and it is therefore truly a second phase of the design exercise. Currently, designers are neither expected to carry out nor paid for such an exercise.

- The design team may interpret the exercise as a critique of their design judgement. This is a difficult problem that is hard to counter unless the original designer plays a part in the activity. The reason given by some value engineers for not including members of the original design team is the danger that old ideas may be defended and their presence may stifle frank comments on the design.
- It is beyond dispute that the value engineering study will effectively take three to four weeks from the design programme; that is, one week prior to the study for the distribution of drawings and information, the study week and the period of time following the study for the submission of the report, discussion and design changes. In some projects this period of time, during which the design will be at a standstill, will be unacceptable. However, in the majority of cases it is capable of being accommodated particularly in view of the fact that the study itself is a watershed between sketch design and working drawings and provides an immoveable date for the completion of the sketch design.
- The resourcing of a study can pose problems associated with the withdrawal of professionals from their home office for a one-week period. It is a condition precedent to a successful study that members of the value engineering team are isolated from their home environment for the study period.

The study therefore is not without its problems but has consistently proved to be a very effective means for the application of value engineering.

Approach 3: The value engineering audit

The value engineering audit is a service offered by value engineers to large corporate companies or government departments to review expenditure proposals put forward by subsidiary companies or regional authorities. The procedures employed follow exactly those of the job plan. Following a proposal the value engineer will visit the subsidiary company or regional authority and undertake a study of the proposal from the point of view of providing the primary functions. The study may be carried out using personnel from the subsidiary company or regional authority or from another company within the group or another regional authority. The study is a global review and normally takes one or two days; it is therefore fast and relatively inexpensive.

Following the review the value engineer will submit a report detailing the primary objective and the most cost effective method for its realisation.

The audit may be criticised on the grounds that it is potentially conflict orientated and that depth is sacrificed for breadth. However, the projects director of one subsidiary company stated that a value engineering audit on one proposal revealed a number of shortcomings with the statement of requirements to the extent that the company now adopts a policy of holding a charette before a proposal is submitted to the parent company.

Approach 4: The contractor's change proposal

The contractor's change proposal (or value engineering change proposal) is a post tender change inspired by the contractor. The US Government include a clause in their conditions of contract which states that contractors are encouraged to submit ideas for reducing costs. If the change is accepted by the design team then the contractor shares in the saving at the rate of 55% of the saving for fixed price contracts and 25% for cost reimbursement contracts. For example, if a contractor on a contract of $250 000 makes a suggestion which the contractor estimates will save $10 000 then, following verification and acceptance by the design team, the contractor will receive $5500. The payment is made by reducing the contract sum by $4500.

The benefit of the clause is that it allows the contractor to be proactive and use construction/engineering knowledge to improve a facility at on-site stage.

The disadvantage of the clause is that the contract may be delayed while the design team investigate the viability of the change. For this reason changes tend to be relatively superficial.

Variations on the formal approaches to value engineering

Although the four approaches detailed above are most often described they are not suitable in every case. The following are applications of the job plan that have been used in practice but do not fall within the standard approaches.

- *The orientation meeting*: Similar to the charette, this meeting is a part of the value engineering procedure operated by New York City, Office of Management and Budget. The meeting of client representatives, design team and independent estimator is held when either the brief or the brief and schematic have been developed. The objectives of the meeting include to provide an opportunity for all taking part to understand project issues and constraints, to give and receive information, and to determine whether all information for a 40-hour study is likely to be available by the date set for the study.
- *The shortened study*: In many cases the estimated project value is lower than the $2–3 million considered to be the lower limit for a full 40-hour study involving a team of six. In this case the 10% rule of thumb is used to

determine how much can be spent on the study. For example, for a project of $500 000 the target savings are $50 000 and the fee for the study $5000. It is now a question of determining how much professional time can be bought for $5000. If a rate of $1200 per person per day is assumed, including expenses, then four person days can be afforded, e.g. three people for one day with one extra day for the value engineer.

- *The concurrent study*: This approach involves the design team themselves coming together on a regular basis under the chairmanship of the value engineer to review design decisions taken. The method has much to commend it in that it answers many of the criticisms levelled at the standard 40-hour study. The extent of involvement of the value engineer needs to be determined in advance so that the fee can be established. The fee bid from members of the design team will also have to account for their extra involvement.

The concurrent study is also suitable for construction engineering contracts in which the design is carried out in stages along with the construction (fast-tracking). The reference point for the comparison of costs was the elemental cost plan in a study of a $100 million office project in Canada, on which a value engineering design meeting was held on site each Wednesday. At initial meetings the functions of the spaces were analysed and five adjacency diagrams generated. These were reduced to three for presentation to the client. For the selected plan a number of structural solutions were generated and analysed on a matrix along with solutions for the electrical and mechanical installations. Once the building form was established, construction work began. Meetings continued with the construction engineer in attendance through to the end of the project. The final cost of the project was $9 million less than the budget.

This brief case study illustrates the importance of demonstrating cost savings in the North American value engineering environment. A disadvantage of the concurrent study is the difficulty of proving the value of the time expended: it is easy to say that the budget was too high.

Conclusion to development in the USA

The development of the value engineering process and the study styles described above reflect the practice of mainstream value engineering in the USA. Development through academic research is relatively limited. Research undertaken by predominantly practitioners tends to concentrate on near market developments of the service often involving enhancements to economic modelling or quality techniques. SAVE International hold an annual conference and publish a quarterly journal.

28 Method and Practice

2.3 The global development of value management

The global development of value management began in the 1960s with the spread of value engineering techniques by North American subsidiary companies to principally Europe and Australia. In the 1970s the manufacturing industries of primarily Japan developed the principles of value engineering and applied them enthusiastically. In the late 1970s and during the 1980s the development of value engineering appeared to be retained within manufacturing companies for competitive advantage and discussion among practitioners decreased. For example, by the end of the 1980s membership of the UK Institute of Value Management fell to below ten. During the 1990s there was a shift from the application of the techniques solely to the design and production of components to the introduction of the principles to the business strategy as a whole. Manufacturing evolved a VM definition to incorporate such notions as an emphasis on the customer and a systematic product and manufacturing development approach, both relating total concept to a customer focused design procedure.

The lever of quality developed by Winston Davies at Jaguar Cars was used to demonstrate graphically that a given quality enhancement can be achieved by applying either minimal effort at the value planning stage or a lot of effort at the value analysis or process stage. This concept is encapsulated in Fig. 2.1.

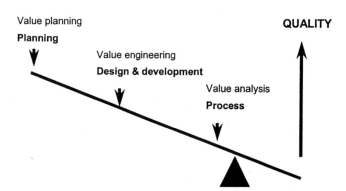

Fig. 2.1 The lever of quality (after Winston Davies).

The lever of quality diagram (Fig. 2.1) influenced thinking incorporated in the lever of value discussed later in this book. Retaining value management as a technique for competitive advantage still exists as illustrated by the total value management methodology outlined by the Ford Motor Company (*Financial Times*, 11 March 2002).

To understand the transference of the value management principle from its original place of conception in the manufacturing industry it is necessary to appreciate the characteristics of value management in a manufacturing context. These are as follows:

- Manufacturing has a long planned development time, although competitive pressures are continually reducing development time.
- Manufacturing has a high level of customer focus.
- Manufacturing has to bear very high product testing costs. This is particularly true in the car industry.
- Manufacturing has high tier one supplier costs, commonly 85% of the final product cost.
- Manufacturing has demonstrated over the past decade a huge willingness to improving efficiency in the face of global competition.
- Manufacturing generally retains design within its own organisation.

Global development in construction followed a fairly similar pattern to that demonstrated in manufacturing. From North American manufacturing, value engineering was introduced into North American construction through the US Navy. The US Navy still retains one of the most active value engineering programmes in terms of both ship production and construction of facilities. Construction value management developed largely through the public sector was reflected earlier in this chapter. The spread to the UK was largely through organisations with North American head offices and the research activities of UK academics. It is interesting to note that a similar process occurred in Australia. During the mid to late 1990s a number of guidance documents were produced in the UK on the subject of value management leading to the conclusion that there might be a turf war amongst construction institutions. This is discussed further in Chapter 9 dealing with the professional development of value management and ethics.

The next section reviews differences in value terminology.

2.4 Value engineering and value management: an overview of terminology and definitions

The terms value management (VM) and value engineering (VE) are those most commonly used in the literature. However, other terms are in use and the full list is as follows:

VM – value management
VE – value engineering
VA – value analysis
VP – value planning
VM – value methodology (the term used in the USA for value management)

The Building Research Establishment (BRE) states that it is not important what the process is called but it has opted to use VM as an all-embracing term for any application.

The Institution of Civil Engineers (ICE) uses VM as an all-embracing procedure. It has also adopted the following terms as sub-processes within VM:

- Value planning (VP) is applied during the concept phase of a project.
- Value engineering (VE) is applied during the definition phase of a project.
- Value reviewing (VR) is applied at any point in the project life cycle to record the effectiveness of the value process which relates to the overall sequence of actions that lead to value improvements.

CIRIA (Construction Industry Research and Information Association) uses VM as an all-embracing term to describe a problem solving technique. However, it states that value management (VM) is aimed at strategy early in the design and that value engineering (VE) is concerned with achieving project objectives cost effectively.

BSRIA (Building Service Research and Information Association) states that VM addresses project objectives at the concept and feasibility stages and that VE addresses engineering issues in the design phases. This seems to concur with the CIRIA document.

The Treasury CUP Guidance Note No.54 *Value Management* makes reference to reviews throughout the life of a project to ensure that the best value for money project progresses through briefing, design and construction. The reviews accord with value and risk opportunities.

The European standard, reviewed in more detail later in this chapter, describes VM as a style of management encompassing various management tools of which value analysis is but one. It does, however, state that VA is a tool used to tackle more tactical issues.

VE is the term favoured historically by the Society of American Value Engineers (SAVE). SAVE has considerable influence internationally in construction and therefore the term is in common use around the globe. The Society of American Value Engineers renamed itself SAVE International in 1977. It now uses the term 'value methodology' as a collective term, highlighting in the SAVE Standard that it includes the processes known as value analysis, value engineering, value management, value control, value improvement and value assurance. The Standard is all embracing. Whilst it does not define the methodology per se, the SAVE Standard embraces an approved job plan, a body of knowledge, typical profiles of value managers and value specialists, and the duties of a value organisation.

The Australian/New Zealand Standard (AS/NZS) reiterates what BRE state, in that VM is synonymous with VE and VA and can be applied any time during a project life cycle.

This book recognises the distinction between VM and VE in construction in the UK as follows:

- Value management (VM) in construction is the term used to describe the total process of enhancing value for a client from a project from the phases of concept through to operation and use. As a process, value management encompasses understanding and providing solutions to the 'business project'. This is the business problem that the organisation needs to solve or address. The business project is unlikely to be defined in construction language but is expressed and communicated in the language of business. For example, the business project may involve reorganisation within the firm that may also require a new facility or refurbishment to a facility as options. As discussed later, it relates directly to the concepts of total quality management.
- Value engineering (VE) is the term used to describe a subset of the value management process, where the focus is on improving value in the design and construction stages of the 'technical project'. This is the manner in which the business project is translated into the requirement for a built facility through design and construction. Value engineering has more application to the concepts of quality assurance, discussed later.

There are many definitions of construction value management but all contain these key ingredients:

- Team orientated.
- Makes explicit the client's value system.
- Functionally driven.
- Proactive and creative.
- Has application at a number of stages during the construction process.

A value management definition is therefore:

> Value management (VM) is a service that maximises the functional value of a project by managing its development from concept to use through the audit of all decisions against a value system determined by the client.

Japanese approaches

Contrary to the approaches of North America, Australia and UK, in Japan value engineering is not an event but rather a continuous process carried out within a philosophy of continuous improvement across all phases of the construction process. The term value engineering is used here as it is the term used by the Society of Japanese Value Engineers. SJVE has strong links with SAVE International to the extent of reciprocity in respect of training structures, qualifications, etc.

The founding structures of value engineering, function definition, evaluation based upon the lowest cost to perform function, the use of FAST diagrams, the job plan, and the use of creativity are essentially the same as for the USA. The focus on elements and components rather than on the form of the project to meet the strategic requirements of the client is also similar to the USA. The differences lie in the continuous process involving the design team and the use of an in-house value engineering facilitator. A function of private sector procurement being design and build means that the contractor carries out value engineering. In public sector projects where design is commonly divorced from construction there are two value engineering opportunities, one with the design team and the other with the contractor.

Differences in culture also impact the approach to value engineering. Whereas the North American approach seeks a demonstrable financial return in a short focused exercise, the Japanese seek satisfaction from a holistic assessment of the problem with a range of possibilities which can be considered long term and determined by consensus amongst the team. The Japanese system is therefore intuitive and future orientated as against the North American system of clear returns now[6].

New South Wales Government total asset management system

The total asset management system TAM2000 is a comprehensive guide to the management of the public sector built infrastructure of New South Wales, Australia. A part of TAM2000 is the Value Management Guideline. This 24-page document issued in January 2001 outlines the value management definition, policy, concept, application, processes and procedures. It describes value management as 'a structured, analytical process for developing innovative, holistic solutions to complex problems' and highlights the characteristics of value management as having a specific methodology based upon creative problem solving involving key stakeholders in a managed team approach to function definition with a focus on value added solutions, integration and project learning.

Applications of value management to government projects are seen to be establishing and verifying project objectives; creating and analysing a range of options for executive consideration; analysing briefs; optimising design solutions; resolving conflicts and improving communications. These applications are repeated in the policy section of the document with emphasis on establishing project objectives, the preparation of the project brief and the consideration of concept and design options. It is also a policy requirement that for projects of A$5 million or over formal value management studies are required and submissions to Treasury should be accompanied by a value management study report and the agency's preferred direction and implementation strategy. There are no formal requirements for projects of less than A$5 million but a value management approach is expected particularly to establishing the project rationale and the consideration of options. Further, the Treasury may require a formal value management action in respect of certain important or sensitive projects.

The guideline is comprehensive and useful in the context of enabling an agency without knowledge of value management to gain a basic understanding. The guideline goes hand in hand with a requirement that NSW Government agencies use approved value management practitioners.

Review of HM Treasury procurement guidance

The key document in the series of ten documents relating to construction procurement is document number 2 (referred to as D2) *Value for Money in Construction Procurement* which refers *inter alia* to Central Unit on Procurement (CUP) guidance number 54 *Value Management*. The former document outlines policy and the latter document gives guidance on procedure.

HM Government procurement policy is to achieve best value for money. D2 states that 'accountability for public funds must not be used as an excuse for missing opportunities to deliver value for money'. The key features of a value for money procurement policy as defined in D2 are:

- Integrating value management and risk management techniques within normal project management.
- Defining the project carefully to meet user needs.
- Taking account of whole life costing.
- Adopting change control procedures.
- Avoiding waste and conflict through team working and partnering arrangements.
- Not appointing consultants and contractors on the basis of lowest initial price alone.

Value management is seen to be a positive focus on value rather than cost and seeks to achieve an optimal balance between time, cost and quality. It is a structured, auditable, accountable, multi-disciplinary process that seeks to maximise the creative potential of all departmental and project participants working together. Departmental project sponsors are encouraged to establish a workable framework for the continuous review of the project as it develops to ensure that it meets departmental needs and objectives, ensuring that the value management plan is drawn up and incorporated into an early draft of the project execution plan.

The generic project flow chart that appears in D2 is adapted as Fig. 2.2 to show the value management and value engineering input points. The 'Gateways' on the diagram refer to the Office of Government Commerce (OGC) Gateway review process.

Review of British Standard BS EN 12973:2000 Value Management

European Standard EN 12973 was approved by CEN on 7 October 1999 and was issued as a British Standard in April 2000. The standard defines value

34　Method and Practice

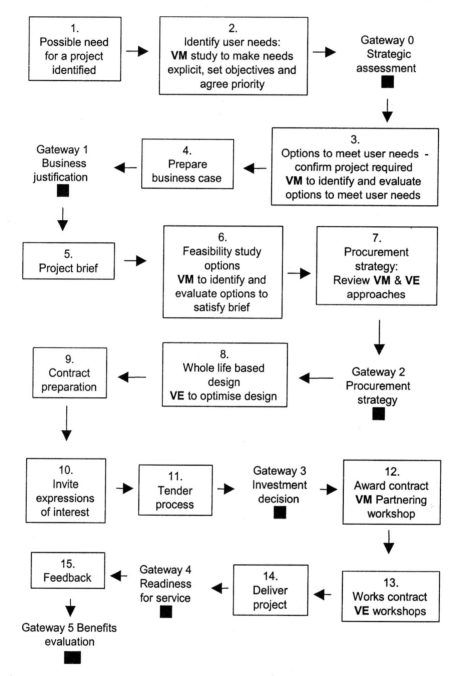

Fig. 2.2　Generic project flowchart (adapted from HM Treasury D2).

management as a style of management particularly designed to motivate people, develop skills and promote synergies and innovation, with the aim of maximising the overall performance of an organisation. Value management provides a new way to use many existing management methods. It is consistent with quality management. It states that value management has been proven effective in a wide range of activities.

The Standard recognises that stakeholders, internal customers and external customers hold differing views of what represents value. The aim of value management is to reconcile these differences while enabling an organisation to achieve the greatest progress towards its stated goals with the use of minimum resources. Value management simultaneously addresses management goals, encourages positive human dynamics, respects internal and external environmental conditions and positively provides the methods and skills for achieving results.

The value management approach outlined in the standard embraces three underlying principles, namely:

- A continuous awareness of value for the organisation, establishing measures or estimates of value, and monitoring and controlling them. In this context value is described as a relationship between satisfaction and the resources used in achieving that satisfaction.
- Attention to the identification of objectives and targets before seeking solutions.
- Maximising innovative and practical outcomes by focusing on function.

Having dealt with the context the remainder of the standard addresses style, functional focus and base techniques. In respect of style the standard highlights the importance of a four-part approach involving teamwork and communication, a focus on what things do rather than what they are (function approach), an atmosphere that encourages creativity and innovation, and a focus on customers' requirements. The functional focus is enhanced by the application of a framework that is applicable to all levels within an organisation, promoting a receptive attitude of mind to the use of functional concepts, methods and tools. Examples of pro forma frameworks for value analysis and the work plan are included in the Standard. The functional focus relates to customer or product needs described as 'use needs' and 'esteem needs'. Use needs, which correlate with user related functions (URFs), are identified as tangible measurable activities. Where these activities involve a product, the product will also relate to the function known as product related functions (PRFs). Esteem needs are the parts of the total need that are subjective, attractive or moral.

The standard defines successful value management as working within a context of human dynamics, methods and skills, management style and environment. The environment described is the environment within which value management operates and takes into account the broader environments of customers, suppliers, statutory and legal constraints and ecological considerations. The methods and skills fall into two categories: the method of undertaking a formal value management study including the value management study plan or agenda, and the methods or techniques used within a value management study. The latter include value analysis, function analysis, function cost, function performance specification, design to cost and design to objectives. Other methods and tools used concurrently are described as creativity, failure mode effects, criticality analysis, life-cycle costs, quality function deployment and many others.

The Standard incorporates some useful information, guidance and frameworks for value management. However, although it may be set out as being sector neutral it does lean heavily towards manufacturing with an emphasis on the product.

2.5 The international benchmarking study of value management

A benchmarking research project, sponsored by the Engineering and Physical Sciences Research Council in the UK, was undertaken during the period 1995–8. It is referenced throughout this book and will be referred to henceforth as the benchmarking study[7]. There was a requirement to set out a reference point or datum around which the benchmarking studies could be conducted and this was a value management methodology developed through research studies conducted by the authors from the mid-1980s onwards.

This section, drawing heavily on work published by the authors through Thomas Telford[8], outlines the developments in the authors' thinking as a background to this book. The section is broken down into two main components. Component 1 comprises the methodology as developed up until the benchmarking study in 1995. The benchmarking study was founded on *Value Management in Design and Construction* published in 1993, which became the benchmark text. Component 2 is that developed as part of the research work from 1995 until publication of the benchmarking study report in 1998.

Component 1: research leading to the benchmark text in 1993

The methodology developed before the benchmarking study was derived from a number of research studies. It has been set out in a series of reports, papers and books.

1986: the first visit to the USA

This first visit to the USA was funded by the Education Trust of the Royal Institution of Chartered Surveyors. It followed from a number of literature reviews on the application of North American value engineering to UK construction. The primary research question to be answered was 'is value engineering another name for the cost planning/control service offered by the UK quantity surveyor?' The research undertaken by the authors took in a wide spectrum of practice in the USA and Canada. They concluded that the service of value engineering was different from QS practice in a number of fundamental areas. It was also highlighted that practice was more diverse than was portrayed in the value engineering texts. The following publications were produced:

- Kelly & Male (1989) *A Study of Value Management and Quantity Surveying Practice*[9].
- Male & Kelly (1989) Organisational responses of public sector clients in Canada to the implementation of value management: lessons for the UK construction industry. *Construction Management and Economics*[10].

1989: the second visit to the USA

The second visit, conducted in 1989, was also funded by the Education Trust of the RICS. The study conducted in 1986 left some questions unanswered: first, identifying the techniques used during early design or orientation phase; second, identifying the specific techniques used for function diagramming; third, identifying techniques used during a 40-hour value management study. During this research study data was gathered in all three research areas and the following publications produced:

- Kelly & Male (1991) *The Practice of Value Management: Enhancing Value or Cutting Cost?*[11]
- Male & Kelly (1991) The economic management of construction projects – an evolving methodology. *Habitat International*[12].

Much of this information is summarised in the earlier part of this chapter.

1991–2: a two-part study of client briefing

Questions relating to the application of value management at the early design or orientation stage still remained unanswered from the 1989 study. The subsequent study was again funded by the Education Trust of the RICS and led to a two-part study. Part 1 was to review the methods of briefing. Part 2 was to propose an alternative approach based upon the techniques of value management. The primary outcome from this research was the recognition of a two-stage briefing process. The first stage dealt with the 'business brief' for a project and the second dealt with the 'technical brief' for the project. A practice manual was produced for the briefing process. It identified in more detail how to conduct a two-stage briefing process using VM. The proposed techniques were tested in an expert seminar hosted by the RICS and attended by representatives of influential client organisations. Three publications were produced:

- Kelly et al. (1992) *The Briefing Process: A Review and Critique*[13].
- Kelly et al. (1993) *Value Management: A Proposed Practice Manual for the Briefing Process*[14].
- Kelly et al. (1993) Functional levels in building design, *Proceedings of CIB W55 Symposium: Economic Evaluation and the Built Environment*[15].

The text that became the forerunner of the benchmark

By the end of 1992 sufficient material had been generated by research for the publication of the book which contained the theories and principles of the North American work and argued the case for a structured approach to VM in a UK construction context. The resulting text was:

- Kelly & Male (1993) *Value Management in Design and Construction*[16].

The 1994 study: VM and strategic project decision taking

Further funding was obtained from the RICS to investigate the application of VM to strategic decision taking on projects. This study was prompted by unanswered questions from the 1991/92 briefing study. North American colleagues, having read the RICS reports, noted that the practice of North American architectural programming was significantly similar to the approach being proposed in the 1993 publication. A further investigation raised a number of issues that were addressed subsequently in the benchmarking study. The 1994 study was recorded in a paper given at the first RICS Construction Research Conference COBRA 1995 and entitled:

- Kelly & Male (1995) Facilities programming. *Proceedings of COBRA'95 RICS Construction and Building Research Conference, Edinburgh*[17].

The pre-benchmarking study methodology up to 1995

Kelly and Male[18] identified four levels where a value study may fit into the project delivery process. The application of value management to building tends to reflect the four-stage process by which building develops. This process can be demonstrated in reverse by considering building characteristics:

- Modern buildings are an assembly of manufactured components.
- The components are joined together to form elements of construction. UK value management consultants find themselves at a large advantage over, for example, their North American counterparts in that the Building Cost Information Service have defined an element as that part construction which performs the same primary function irrespective of the components from which it is made. Therefore in the UK the cost of function provision at an elemental level can be easily determined. Value engineering is the most

appropriate activity for determining the value for money of elements and components.
- The elements go together to surround and service space. In considering space it is necessary to reflect on the relationship between the function of the space and the cost of space. The latter is entirely a function of the cost of the elements that surround and service the space. Space in buildings is an unusual concept because, although it is the space that is used, it is inherently free; like air, it is fundamental to life but has no cost. However, elements to physically and environmentally support space are required in order, for example, to carry out office functions in peace and comfort 100 metres above a busy street.
- Designated spaces together reflect the corporate organisation and business strategy of the client. Value management techniques are used to ensure that the building and spaces achieve a strategic fit with the requirements of the client at a cost that represents best value for money.

The levels are detailed below and in Fig. 2.3. The levels are:

- Level 1 Concept
- Level 2 Spaces
- Level 3 Elements
- Level 4 Components

The four levels usefully form the basis for the study styles described in Chapter 5. It was thought that these levels would have a direct impact on how a value management study is carried out and they represent the demarcation between the main focus of value management and value engineering.

Study styles

Nine different value management study opportunities were identified by Kelly & Male[19]. The choice of approach depended on the client problem to be addressed and the stage of the project life cycle that was to be targeted for study. It was seen that, once a study approach had been appropriately selected, certain tools and techniques would be applicable to that study. Five basic approaches to undertaking a value management study were identified and are illustrated in Fig. 2.4. These were termed:

(1) *VM1*, where the personnel involved are a study facilitator and client representatives who have an input into the definition of the project. The primary task of this team was seen as to identify the project task, client needs and wants, stated explicitly in function terms. They would be primarily concerned with level 1 but might address level 2 issues. The aim of this study would be to ensure that the client requires a built solution.
(2) *VM2*, where the personnel are a study facilitator and client representatives

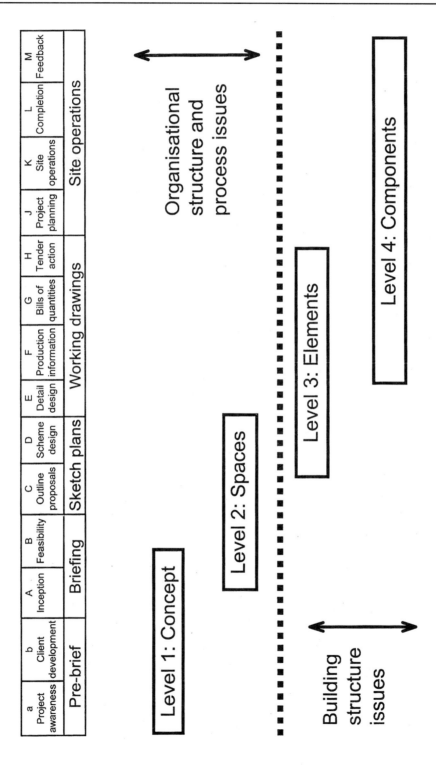

Fig. 2.3 Decision levels and the project life cycle.

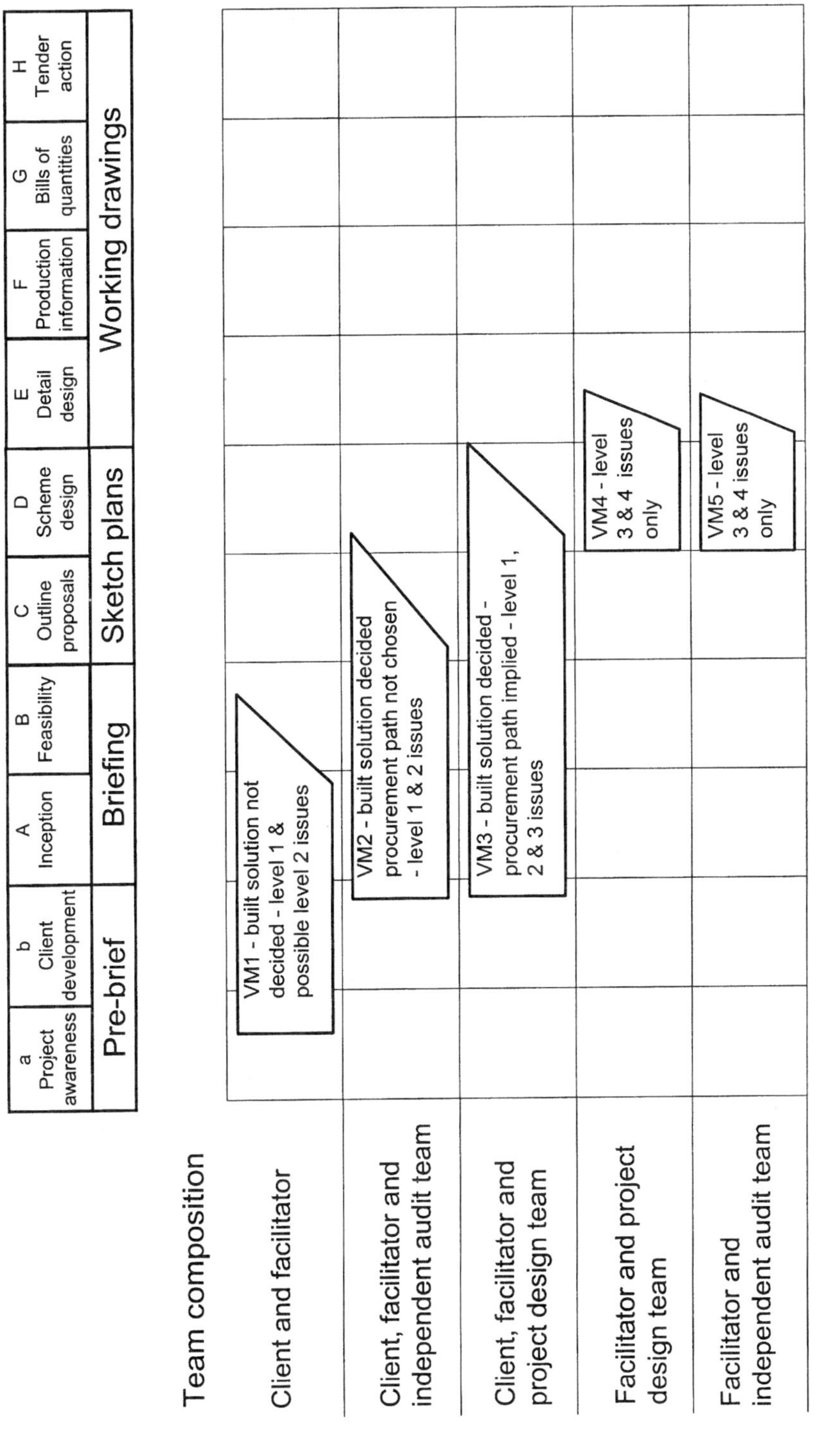

Developments in Value Thinking   41

a Project awareness	b Client development	A Inception	B Feasibility	C Outline proposals	D Scheme design	E Detail design	F Production information	G Bills of quantities	H Tender action
Pre-brief		Briefing		Sketch plans			Working drawings		

Team composition

Client and facilitator — VM1 - built solution not decided - level 1 & possible level 2 issues

Client, facilitator and independent audit team — VM2 - built solution decided procurement path not chosen - level 1 & 2 issues

Client, facilitator and project design team — VM3 - built solution decided - procurement path implied - level 1, 2 & 3 issues

Facilitator and project design team — VM4 - level 3 & 4 issues only

Facilitator and independent audit team — VM5 - level 3 & 4 issues only

Fig. 2.4 The five basic approaches to value management.

together with an independent audit design team. This design team might not be involved with the project design. At this point, the decision to build was seen as having been taken but the procurement route would still be open. The primary task of this team was seen as to address level 1 and level 2 issues. This allowed the procurement route to be the subject of investigation when strategic project issues were explored with the client. It would also lead to faster procurement since the project could be appraised functionally and technically before the brief was detailed for design.

(3) *VM3*, where the personnel are the study facilitator and client representatives together with the project design team. At this point the decision to build was seen to have been taken and the appointment of a project team would indicate a traditional procurement route but with the options for construction management or management contracting still open. This type of study was seen to be non-conflict orientated and with increased possibilities for implementation. The exercise could be used for briefing the design team and maximises project knowledge. This exercise was seen to be inexpensive since the design team is not likely to charge an additional fee for this briefing exercise.

(4) *VM4*, where the personnel involved are a study facilitator together with the project design team. The project was seen to have reached sketch design. While the team would address level 2 issues, in the absence of a client representative they would be more likely to be concerned with levels 3 and 4. The absence of the client would indicate that the client's value system was embedded in the design to date. This study style was seen as non-conflict orientated but could increase the design period, particularly where reference back to the client was required.

(5) *VM5*, where the personnel involved are a study facilitator together with an independent audit design team. The project was seen to have reached sketch design. While the team would address level 2 issues, in the absence of a client representative they would be more likely to be concerned with levels 3 and 4. The absence of the client would indicate that the client's value system was embedded in the design to date. This provides objectivity and the opportunity for increased technology transfer. However, it was seen that there was an increased risk of an adversarial relationship developing that could increase the design period.

Four basic approaches to undertaking a structured cost study were identified and are illustrated in Fig. 2.5. These were termed:

(1) *SCS1*, where the personnel are a study facilitator and client representatives together with an independent audit design team. This design team was seen as not being involved with the project design. The project would have reached sketch design stage. This provides objectivity and the opportunity for technology transfer. However, adversarial relationships with the existing design team are a possibility. The primary task of this team was seen as

addressing level 3 and 4 issues. However, there might be small changes to level 2 issues.

(2) *SCS2*, where the personnel are a study facilitator and client representatives together with the project design team. The project would have reached sketch design. This was seen as non-conflict orientated but the incentive for change is limited. The primary task of this team would be to address level 3 and level 4 issues. However, there might be small changes to level 2 issues.

(3) *SCS3*, where the personnel involved are a study facilitator together with the project design team. The project would have reached sketch design and the team will address level 3 and 4 issues. The absence of the client would indicate that the client's value system is embedded in the design to date. This is non-conflict orientated but the incentive for change is limited.

(4) *SCS4*, where the personnel involved are a study facilitator together with an independent audit design team. The project would have reached sketch design and, while the team might address level 2 issues, in the absence of a client representative the team would be more likely to be concerned with levels 3 and 4. The absence of the client would indicate that the client's value system is embedded in the design to date. This was seen as potentially adversarial but is objective and with the opportunity for technology transfer. However, the implementation rate of potential savings could be low and there is a danger that quality could be compromised. A quantity surveying cost study is most likely to address level 3 and 4 issues as illustrated following the completion of the sketch design and the cost plan.

From Fig. 2.5 it was postulated at the time that the methodology could be summarised as:

- Associating value management with the management of projects, linking the latter and corporate and business strategy. Value management is seen as a business term that describes a philosophy, service and approach that ensures a 'value thread' is maintained throughout the project life cycle. It achieves this through VM or cost studies as points of convergence in the project where there is a need to brings different skills and knowledge together to ensure value for money.
- Identifying that both value management and cost management studies could be undertaken, with value management studies occurring earlier in the project life cycle.
- Defining points for value opportunities in project terms with different levels for a study and identifying the characteristics of that study.
- Identifying certain inputs and outputs for each of the value management and cost management opportunities.
- Identifying two stages to the briefing process, resulting in a policy brief and concept brief as the outputs.

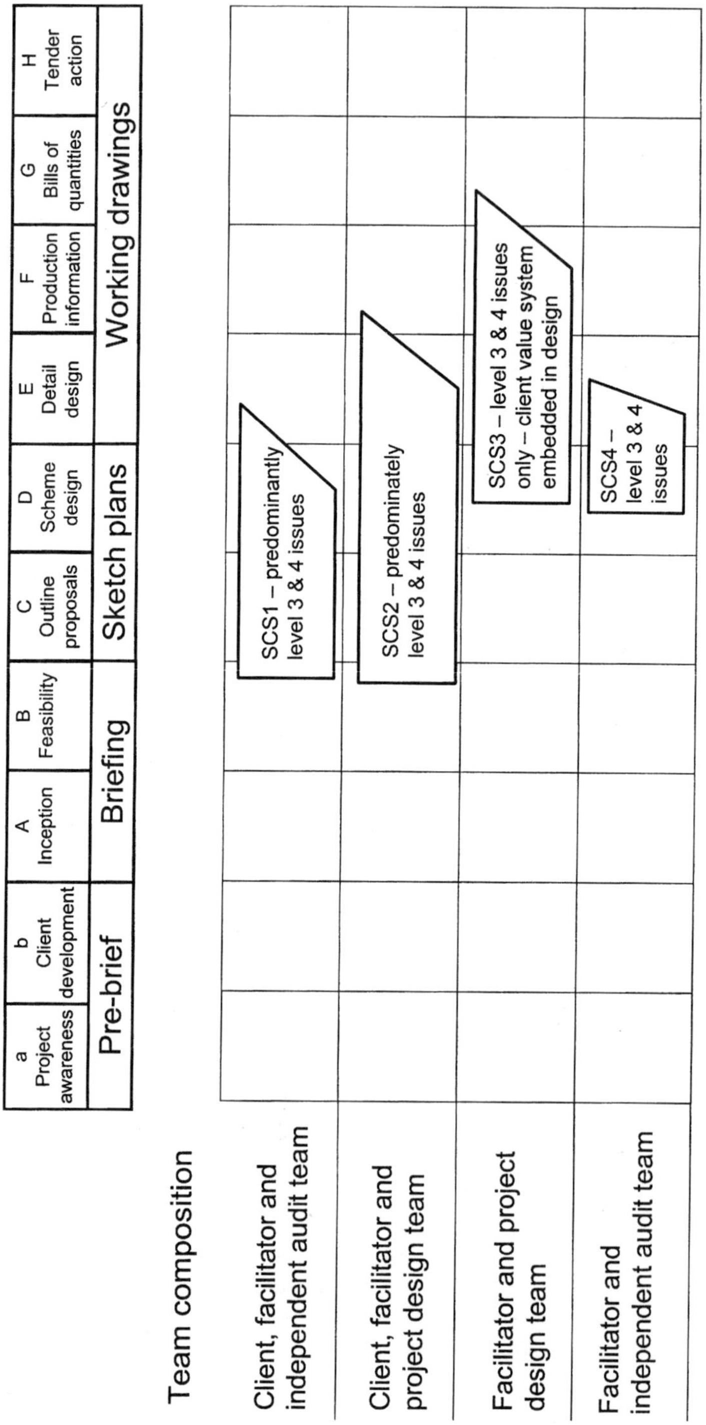

44 Method and Practice

a Project awareness	b Client development	A Inception	B Feasibility	C Outline proposals	D Scheme design	E Detail design	F Production information	G Bills of quantities	H Tender action
Pre-brief		Briefing		Sketch plans			Working drawings		

- SCS1 – predominantly level 3 & 4 issues
- SCS2 – predominately level 3 & 4 issues
- SCS3 – level 3 & 4 issues only – client value system embedded in design
- SCS4 – level 3 & 4 issues

Team composition

Client, facilitator and independent audit team

Client, facilitator and project design team

Facilitator and project design team

Facilitator and independent audit team

Fig. 2.5 The four basic approaches to cost management.

Component 2: the 1995 EPSRC IMI International Benchmarking Study

In order to establish a datum for the research project, the authors made explicit their methodology. This has been consolidated schematically into the datum for the benchmarking exercise and is set out in Fig. 2.6. The methodology was supported by an in-depth review of live value management studies conducted by the authors. This revealed that the actual tools and techniques applied in a study depend on a number of factors, including:

- The nature of the client and the motives for carrying out the study.
- The type of project and its stage in the design process.
- The composition of the VM team.
- The practical constraints imposed on the study itself.

From this, it was identified that the process of value management could be broken down into three main areas: the *inputs* required for a study (now termed the orientation and diagnostic phase); the *process* of value management during a workshop (now termed the workshop phase), and the *outputs* of the value management workshop (now termed the implementation phase). At each stage of the study it was identified that a number of steps are undertaken. For each step there are a number of techniques available, the most appropriate depending upon the circumstances.

Figure 2.6 shows these steps in the shaded boxes under the various job plan stages, with a selection of techniques that had been employed in practice. It was not the intention to describe the 'classic' job plan stages but the schematic was used as the basis of the detailed dialogues with benchmarking collaborators on the research project.

The elicitation of the methodology into an explicit framework indicated a number of issues for further consideration:

- The information stage is often cited as the most important stage in the value management process. It lays the foundations on which the remainder of the exercise is built. The process diagram shows that the information stage has the greatest number of possible activities associated with it, particularly when pre-workshop information is taken into account.
- It is notable that there is little or no variation of techniques applied at the creativity stage, although the literature reveals that there are a fair number of creativity techniques[20], with brainstorming proving to be by far the most popular in practice.
- Although Figure 2.6 shows the process as a sequential series of well-defined steps, the practice of value management was not so clear cut. There is overlapping and iterating of the team's activities.
- The exploration of the methodology also highlighted that the concept of levels is robust regardless the type of project – civil engineering, building, process plant, etc. However, civil engineering and process plant studies were

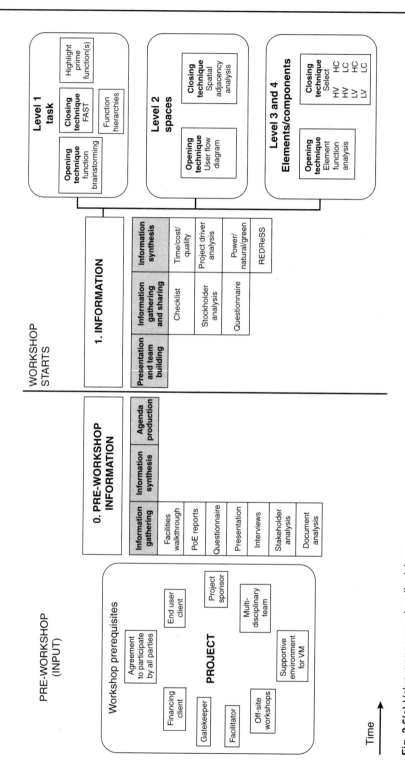

Fig. 2.6(a) Value management methodology.
Source: Adapted from Fig. 5, Male et al. (1998) *The Value Management Benchmark*[21].

Developments in Value Thinking 47

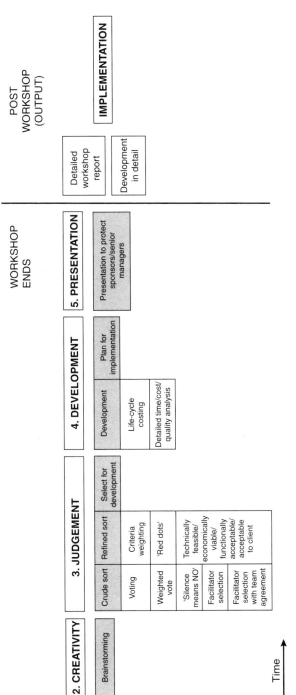

Fig. 2.6(b) Value management methodology.
Source: Adapted from Fig. 5, Male *et al.* (1998) *The Value Management Benchmark*[21].

similar in that studies were normally addressing levels 1, 2 and/or 3 whereas workshops on building projects encompassed levels 1 to 4.

This provided the foundation for the benchmarking study[22] published in 1998 by Thomas Telford, and referred to throughout the remainder of the text.

2.6 Conclusion

From the time when Miles first structured value thinking into a practical service in 1947 developments have been taking place. The job plan, although widely adapted, retains the fundamental elements of Miles's original job plan described in this chapter. Retained in manufacturing for many years, value engineering migrated to the construction industry in the late 1960s and spread through the industrialised world. A number of countries have value institutions that play a greater or lesser role in the establishment of standards and certification of training.

In the UK value management in construction evolved in the late 1980s. The first UK textbook, *Value Management in Design and Construction* by the authors, was published in 1993[23]. Since that time value management in construction has evolved to become an established service with commonly understood tools, techniques and styles. The styles have majored on structured workshops whose agenda has developed from the structures outlined in the USA value engineering workshops described in this chapter enhanced through international benchmarking and ongoing action research in value management. These latest processes and study styles are described in Chapter 5.

The authors have concluded that value management is a service with three primary core elements that distinguish it from any other management service; first, a value system; second, a team-based process; and third, the use of function analysis to promote in-depth understanding. The concept of value is discussed further in Part 2 of this book. Function analysis, teams, team dynamics and facilitation are described in the following chapters.

2.7 References

1. Miles, L. D. (1989) *Techniques of Value Analysis and Engineering*, 3rd edn. Lawrence D. Miles Value Foundation, Washington D. C.
2. Kelly, J. & Male, S. (1993) *Value Management in Design and Construction*, E. & F. N. Spon, London.
3. Norton, B. R. & McElligot, W. C. (1995) *Value Management in Construction: A Practical Guide*. Macmillan, Basingstoke.
4. Fallon, C. (1980) *Value Analysis*, 2nd edn. Lawrence D. Miles Value Foundation, Washington D. C.

5. Kelly, J. & Male, S. (1993) *Value Management in Design and Construction*, E. & F. N. Spon, London.
6. McGeorge, D. & Palmer, A. (1997) *Construction Management: New Directions*. Blackwell Science, Oxford.
7, 8. Male, S., Kelly, J., Fernie, S., Gronqvist, M. & Bowles, G. (1998) *The Value Management Benchmark: A Good Practice Framework for Clients and Practitioners*. Published Report for the EPSRC IMI Contract. Thomas Telford, London.
9. Kelly, J. R. & Male, S. P. (1989) *A Study of Value Management and Quantity Surveying Practice*. Surveyors Publications, London.
10. Male, S. & Kelly, J. (1989) Organisational responses of public sector clients in Canada to the implementation of value management: lessons for the UK construction industry. *Construction Management and Economics*, **7**, No. 3, 203–216.
11. Kelly, J. R. & Male, S. P. (1991) *The Practice of Value Management: Enhancing Value or Cutting Cost?* Royal Institution of Chartered Surveyors, London.
12. Male, S. & Kelly, J. (1991) The economic management of construction projects – an evolving methodology. *Habitat International*, **14 2/3**, 73–81.
13. Kelly, J., Macpherson, S. & Male, S. (1992) *The Briefing Process: A Review and Critique*, Royal Institution of Chartered Surveyors, London.
14. Kelly, J., Male, S. & Macpherson, S. (1993) *Value Management: A Proposed Practice Manual for the Briefing Process*, Royal Institution of Chartered Surveyors, London.
15. Kelly, J., MacPherson, S. & Male, S. (1993) *Functional levels in building design*, Proceedings of CIB W55 Symposium: Economic Evaluation and the Built Environment, Lisbon, **4**, 115–25.
16. Kelly, J. & Male, S. (1993) *Value Management in Design and Construction: The Economic Management of Projects*. E. & F. N. Spon, London.
17. Kelly, J. & Male, S. (1995) Facilities programming. *Proceeding of COBRA '95*. Construction and Building Research Conference, Edinburgh, September 1995, **1**, 99–106.
18. Kelly, J. & Male, S. (1993) *Value Management in Design and Construction*, E. & F. N. Spon, London.
19. Kelly, J. & Male, S. (1993) *Value Management in Design and Construction*, E. & F. N. Spon, London.
20. Parker, D. E. (1985). *Value Engineering Theory*. McGraw Hill, New York.

21, 22. Male, S., Kelly, J., Fernie, S., Gronqvist, M. & Bowles, G. (1998) *The Value Management Benchmark: A Good Practice Framework for Clients and Practitioners*. Published Report for the EPSRC IMI Contract. Thomas Telford, London.
23. Kelly, J. & Male, S. (1993) *Value Management in Design and Construction*, E. & F. N. Spon, London.

3 Function Analysis

3.1 Introduction

Function is defined in Chapter 2 as a characteristic activity or action for which a thing is specifically fitted or used or for which something exists. Therefore something can be termed functional when it is designed primarily in accordance with the requirements of use rather than primarily in accordance with fashion, taste or even rules or regulations. A service (education, banking, insurance, etc.) is functional when all the energy input into the service is focused 100% on the customer; a project is functional when its output contributes 100% to core client business; and an element or component is functional when its purpose in the context of the whole is 100% efficient.

As stated in Chapter 1, strategic and organisational issues are the domains of value management while the technical space, element and component issues are the preserve of value engineering. In this chapter strategic function analysis is described together with a functional approach to elements and components. The chapter is a synthesis of literature together with data derived from action research studies. The aim of the chapter is to describe the process of function analysis, discussion being restricted to the activity of function analysis only. The developments described in the previous chapter are drawn together with the data in this chapter to form a holistic approach to the value management service, which is developed and discussed in Chapter 5.

3.2 Strategic function analysis: the mission of the project

A project is defined as 'the investment by an organisation to achieve an objective within a programmed time that returns added value to the business activity of the organisation'. This is a simplistic definition discussed in more detail in Part 2 but one which is correct for the purposes of this chapter at the strategic stage. In a value management study it is important to:

- Analyse all available information and be fully aware of the problems which lie at the commencement of the project.

- Define the project's function or mission as a simple, clear and understandable sentence.
- Make overt the client's value system.

The client's value system is described in detail in Chapter 8. This section addresses the first two bullet points above, namely to discover all of the necessary information and to define the project's mission.

At the strategic stage it is fundamentally important to analyse all of the information on the issues that surround the project. Ideally the aim would be to discover facts, casting aside anything that did not abide by some physical law or was not capable of significant corroboration. The dangers of not doing this are clear. President Thomas Jefferson writing in 1801 stated, 'to seek out the best through the whole Union we must resort to other information, which, from the best of men, acting disinterestedly and with the purest motives, is sometimes incorrect'. Another relevant but non-attributed quotation is, 'no decision can rise above the quality of the information upon which it was based'. Unfortunately, the majority of projects are based on information discovered through expert testimony or pure gut feeling, and while this is unsatisfactory there is also a danger of spending unjustified time in information seeking in order to be 100% correct. A decision therefore has to be taken on the relative risks of acting on information that could be incorrect or incomplete.

Good quality information lies at the heart of the correct definition of any project, the awareness of which will generally fall within the responsibility of a client organisation. The project awareness stage may lead to a development stage within the client organisation, which is the key stage to implement value management. At the development stage a project sponsor/manager may be appointed to add detail to the basic strategy. It is assumed here that the project sponsor/manager (referred to hereafter as PM) will recommend the appointment of a value management facilitator to organise and run value management workshops. Procedures for the first workshop undertaken at the strategic development stage are described below. The following procedures have been researched in detail and tested in practice. It is important to focus on the fact that the project currently exists in terms of 'the investment by an organisation to achieve an objective within a programmed time that returns added value to the business activity of the organisation'. At this stage the need for a construction project to satisfy a corporate objective has yet to be confirmed.

Stage 1: orientation and diagnostic phase – plan strategic information discovery

The facilitator will first undertake an orientation exercise either through interviews or by holding an orientation workshop. Pre-workshop interviews can be undertaken with a number of objectives:

- To focus the facilitator, PM and client team on the strategy.
- To introduce the facilitator to the key stakeholders.

- To brief key stakeholders on the aims of the value management workshop.
- To outline the likely approach to the project concerned.
- For the facilitator to compile briefing documents for circulation to workshop participants ahead of the workshop.
- For the facilitator to compile an agenda for circulation to participants prior to the workshop.
- For the facilitator to ascertain which of the key stakeholders has the primary information, influence, and executive authority. This will permit the facilitator to compile the workshop participants' list, often in discussion with the client.

This stage may also be undertaken at an orientation workshop but recognising the identification of absent stakeholders with primary information is difficult and hidden agenda are more likely to remain hidden.

The facilitator can work with the client in selecting the members of the value management team or can request that the client build the membership of the team based on the inclusion of those stakeholders with an input relevant to the strategic stage in the development of the project. The ACID test, described in Chapter 8, is used to determine who should be a member of the team. Generally team membership is greater in number at the strategic stage of projects when a large number of issues are being considered and smaller when the technical details of the project are being investigated.

Factors which the facilitator and/or project manager may wish to take into account in selecting the team, are:

(1) Limit multiple representations from one organisation or one department. For example, three members from one organisation or one department where other organisations and departments have single representation will lead to a weight of argument in favour of the multi-represented organisation or department.
(2) Understand the hierarchical mix (senior, subordinate) within the team.
(3) Understand the relationships between team members; for example, one member of the team may be dependent financially on another.
(4) Consider the completeness of the team. Discuss with the client any apparently missing members.

Stage 2: workshop phase – strategic information discovery

Morris & Hough[1] in their study of major construction projects identified that the probability of failure was high when certain key information was missing. In their study this failure could be attributed to a number of identifiable factors. It appears logical that if this information were present then the probability of failure would be diminished. The following have been adapted from the Morris and Hough list.

(1) *Organisation*: The client's business is identified together with the place of the project within the business and the users of the project (who may not necessarily be a part of the client organisation). Under this heading there would be an investigation of the client's hierarchical organisational structure, and the key activities and processes that would impact the project. It must be recognised that most organisations are dynamic and therefore subject to continual change, which should be recognised and recorded. In addition it is important to record the decision making structures of the client and the way in which these will impact the project. These decision making structures become more important in situations where a single project sponsor or project manager represents the project team. The limits to the executive power of the project's sponsor or project manager should be clearly defined. Following the discussion of the organisational structures it should be possible to identify all those who have a stake in the project. Stakeholders should be listed and their relative influence assessed.

(2) *Context*: The context of the project should recognise such factors as culture, tradition or social aspects. Cultural aspects may include the relationship of one department with another or the fitting out and general quality of the environment, e.g. a courthouse. Tradition can cover such aspects as corporate identity, which may be important in such areas as retailing. Social aspects will generally relate to the provisions made by the client for the workforce such as dining and recreation facilities, sports and social club activities, crèche etc.

(3) *Location*: The location factors will relate to the current site of the project, proposed sites or the characteristics of a preferred site where the site has not to date been acquired.

(4) *Community*: It is important to identify the community groups who may require to be consulted with respect to the proposed project. Some market research may have to be undertaken to ascertain local perceptions. The positioning of the project within the local community should also be completely understood.

(5) *Politics*: The political situation in which the project is to be conceived should be fully investigated through the analysis of local government and central government policies and client organisational politics. The latter is often difficult to make overt at a workshop of representatives of different client departments; however, client politics are a key driver behind any project.

(6) *Finance*: The financial structuring of the project should be determined by considering the source of funding, the allocation of funding and the effects of the project cashflow on the cashflow of the client organisation. The latter is particularly important when dealing with organisations working with annual budgets.

(7) *Time*: Under this heading are the general considerations regarding the timing of the project including a list of the chronological procedures that

must be observed in order to correctly launch the project. In situations where the project is to be phased, time constraints for each stage of the project should be recorded. This data becomes the basis of the construction of a time line diagram described below.

(8) *Legal and contractual issues*: All factors that have a legal bearing on the project are listed under this heading including the extent to which the client is risk averse and requires cost certainty. Also included here is data relating to the client's partnership agreements with suppliers and contractors.

(9) *Project parameters and constraints*: A primary objective of the value management workshop at the strategic briefing stage is to fix the primary objectives of the project. Therefore, it is important that the team understand that the workshop is the end of one stage in the development of the project. Discussions must take place on the evolution of the project to the time of the workshop and to measure the extent to which key stakeholders believe that the project is still evolving. Any constraints surrounding the development of the project should be discussed and recorded.

(10) *Project drivers*: A project driver is defined here as any factor which gives impetus to the project. The factor may be legislation, a champion or a change within the organisation. In some situations drivers may be nested and become confusing. For example, an internal enclosed room is to be created within an existing naturally ventilated building. The room requires mechanical ventilation. The installation of the mechanical ventilation poses problems and begins to drive the design. All effort is then transferred to solving the problem of mechanical ventilation, which is seen as the driver.

(11) *Change management*: As discussed in detail in Chapter 8, to ignore change management is to prejudice the project. Issues of change management must be explored and the extent of control, flexibility and risk appreciated.

Working with the team – issues analysis

The method described here is typical of a value management workshop at the strategic stage of projects. At the commencement of the workshop the facilitator will ask the team for all factors impacting the project. The issues will tend to be uncovered in a relatively random fashion and therefore the facilitator records the issues on repositional sticky notes. After the team has exhausted all of the issues impacting the project – typically about half an hour to one hour – the team are asked to the sort the sticky notes under the 11 generic headings above on a master issues sheet. Where a particular checklist heading has no notes attached to it, the facilitator will ask for any issues under that heading. It is common for additional headings to be uncovered at this stage and these too are defined on card.

An alternative to the above procedure is for the facilitator to request issues under each heading in turn. This tends to ensure that each heading is addressed but can stifle the introduction of new headings.

When involved with large team facilitation or when the full workshop team has superior–subordinate or employer–contractor members it may be necessary to split into smaller teams for the brainstorming of issues. Ideally in this situation it would be preferable to have a facilitator per group but practically and logistically it is probably necessary to ask a team member to act as facilitator for their small group. Splitting into small groups ensures that no one is constrained in the presence of their superior or employer and in this situation hidden agenda tend to be exposed. At the end of brainstorming all the issues from all teams are put under headings on a master issues sheet. Invariably there will be duplicate issues and these are removed.

Once the master issues sheet is complete the facilitator will read the issues back to the team. Any queries as to meaning are answered by the team and any extra issues added. Where an issue is thought to have been incorrectly attributed to a heading it is moved or where the team believe that the issue should appear under more than one heading multiple sticky notes are written.

Prioritisation of issues 1

Brainstorming issues can easily result in over 100 issues being raised. It is impossible within a reasonable time to focus on all issues and a means of reducing the issues to those of high importance is required. The team is therefore asked to undertake a prioritisation exercise. Before commencing the exercise it is necessary to determine the basis that individuals are using for their own prioritisation. There is a strong argument to suggest that the client value system should be the basis for prioritisation and therefore the next stage in the process is to establish the client value system paired comparison described in Chapter 8.

Client value system

Undertaking the client value system exercise at this point takes the attention of the team away from the project and refocuses on the place of the project within the client's business. With the knowledge of the client's value system the team can once again address the issues.

Prioritisation of issues 2

Once the client value system paired comparison have been agreed and displayed the team again turns its attention back to the issues. The facilitator will hand to each team member some black sticky dots, in number approximately 10% of the number of issues on the master sheet, i.e. if there are 100 issues on the master sheet each team member is given ten dots. The dots are spent on those issues that the particular team member believes important bearing in mind the client's value system. On completion of this exercise each team member is given three to five red sticky dots to spend on those issues that already have black dots and that the team member believes are so important that the project may not move unless that issue is resolved.

Investigating the high priority issues

At this stage all of the issues are under headings and ranked by importance. The facilitator then asks for more information on the high priority issues i.e. those denoted by multiple black dots or more than one red dot. Further information is recorded on the flip chart. It is common for some issues to be ranked highly important by the majority of the team. The issues analysis, client's value system and the ensuing information on key issues is pinned to the wall and surrounds the team for the remainder of the workshop. It is a rule of the workshop that what is displayed on the wall is agreed. There is plenty of opportunity for any member of the team to raise disquiet about any point either in team forum or in confidence with the facilitator during a break.

Working with the team: information review

A review of all information is assisted by the use of:

- A simple timeline
- A project driver's analysis
- The REDReSS technique

A *simple timeline* can be drawn on a flip chart with commencement dates, completion dates and major milestone events. This is an appropriate stage to ensure that all members of the team are in agreement as to what the various terms mean. For example, completion can mean:

- Practical completion – complete for all practical purposes and the client can take possession although some minor external works and snagging may need to be completed.
- Fully complete – the facility has no further building work to be carried out and commissioning of all of the services systems is complete.
- Operational – the client fit out is complete and the building is fully functional.

Anecdotal evidence abounds regarding multiple interpretations of the word 'complete' by different members of a value management team. The time difference between practical completion and operational on some projects can be measured in months rather than weeks.

A *project driver's analysis* might be required where this has not been addressed within the issues analysis.

REDReSS is an acronym for the final stage of the information validation exercise. Key prompts allow a final analysis of the information to ensure that it precisely represents the project.

- **R**eorganisation. A final check is made to determine the extent to which the client is likely to reorganise during the project's life cycle. In the event that a decision to build is taken, the question becomes one of determining the extent to which the client is likely to reorganise during the design and construction period and/or immediately thereafter. It is tempting to address this problem by maximising flexibility in the solution; however, it should be recognised that this contains a cost penalty. There are a number of examples of where clients have briefed based upon the present and have ignored the future even though that future could reasonably be ascertained.
- **E**xpansion. This question invites an assessment of whether the project will remain fixed or whether a further phase is likely. In the event that a decision to build is taken the question becomes one of: 'what expansion is likely and how can it be accommodated in terms of space and services within the existing and/or planned operating infrastructure'.
- **D**isposal. A construction project will have a predetermined life whether through economic redundancy or physical failure necessitating refurbishment or reconstruction. When this stage is reached, some disposal activity will occur. The question posed here is whether anything worthwhile can be incorporated into the design of the project which will assist 'disposal'.
- **Re**furbishment and maintenance. What are the client's refurbishment and maintenance policies? Can anything be factored into design thinking which will assist those policies? Examples of policies could be that maintenance has to be carried out in a secure environment at predetermined times. This may put pressure on reliability and redundancy issues.
- **S**afety. All projects will have to comply with the requirements of health and safety legislation. However, some clients may wish to extend the issue of safety for instance so that it is clearly demonstrable rather than just compliant.
- **S**ecurity. Some facilities will attract higher security than others. This factor has to be expressed in a manner such that it can be measured. Again consideration of the future might allow security systems to be easily incorporated.

The aim of the review is to finally capture any missing information and to sensitise the team for the function analysis exercise. Information discovery is an opening exercise that relies on information being presented in a manner entirely untainted by interest, motives and hidden agenda. This is achieved through balances and checks being present within the team. The following stage uses the information to derive functions in a closing exercise as illustrated in Fig. 3.1.

Function Analysis 59

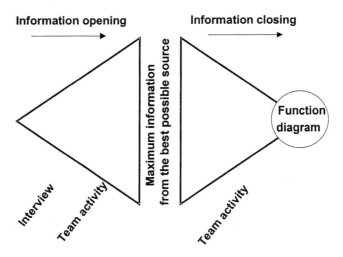

Fig. 3.1 Information activity.

3.3 Strategies, programmes and projects

The issues analysis will highlight the complexity of projects. Often projects are nested within programmes which themselves make up a strategy, a concept illustrated in Fig. 3.2. The strategic cascade is a useful way of determining the place of projects in a situation which is complex. Consider the following example.

An English local authority has purchased a 20 square mile hill site from the crown estate to form a country park adjacent to a large industrial town. The authority intends to undertake limited forestry and farm deer and trout as a part of a job creation scheme. The authority is also keen to promote conservation

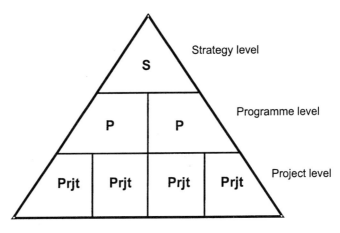

Fig. 3.2 Strategic cascade.

particularly within the town's schools. An established campsite exists but there is a tradition of wild camping in the area. The authority has instituted a project to build a visitor centre at the site with facilities, craft shops and a restaurant. The issues analysis has highlighted the potential for conflict with neighbouring landowners and also the fact that any commercial development at the visitor centre will compete with local shops. However, it is recognised by many that the authority will become a significant employer in the area. The site chosen for the visitor centre is 1 mile from an established middle-class community on the outskirts of the town but is separated from it by a river.

Application of the following technique demonstrated that the project was not simply one of building a visitor centre.

3.4 Function diagramming

Function diagramming at the strategic level

This section introduces the concept of functional analysis and its use at the strategic briefing stage of a project to derive the project's mission through function diagramming. The technique of function diagramming is attributed to a VE practitioner Charles Bytheway[2] who gave it the acronym FAST (function analysis system technique). At this stage the solution to the primary requirements of the project has not yet been formalised and therefore the project is defined simply as 'the investment of resources for return'. Indeed one of the objectives of the use of functional analysis at the strategic briefing stage is to lay the foundation for the solution to the project problems that offers the best value for money. The functional analysis technique relies upon the discovery of all relevant information through the issues analysis and the structuring of that information in a way that leads to the recognition of the primary objective of the project. Function diagramming is a closing technique as illustrated in Fig. 3.1. The facilitator's task, once all of the information is available, is to focus the team on the prime function of the project. The closing technique begins with a brainstorming exercise concerned with 'What exactly are we trying to do here?'

The team brainstorm functions that are required by the project. These functions may be high order executive functions or relatively low order wants. All functions are expressed as an active verb plus a descriptive noun, and are recorded on sticky notes and scattered randomly across a large sheet of paper. The facilitator will continually prompt the team to generate functions by referring back to the information from the issues analysis, timeline and REDReSS.

At the completion of the brainstorming session the team are invited to sort the notes into a more organised form by putting the highest order needs into the top left-hand corner of the paper and at the lowest order wants into the bottom right-hand corner. All other functional descriptions are fixed between these two extremes. It should be emphasised to the team that this is an iterative process and therefore any team member is entitled to move a previously ordered sticky note.

Although this sounds confrontational it is very rare for disagreement to occur and ultimately the correct ordering of all of the functions is achieved.

How the next phase is undertaken is the subject of some debate. Some facilitators will allow the team to construct the diagram in its entirety. However, this can be difficult where there is a large team. The preferences are for the facilitator and perhaps one or two team members to delay going to lunch and to complete the diagram. Later discussion can take place on whether the diagram and particularly the prime function is correct.

The form of the diagram will depend largely upon the project under review. The references give a background to function diagramming. The section which follows gives a description of three diagramming techniques which have been found to work in practice.

Generation of functions

The functions were generated by the team in a random fashion on sticky notes and placed on a large sheet of paper as illustrated in Fig. 3.3.

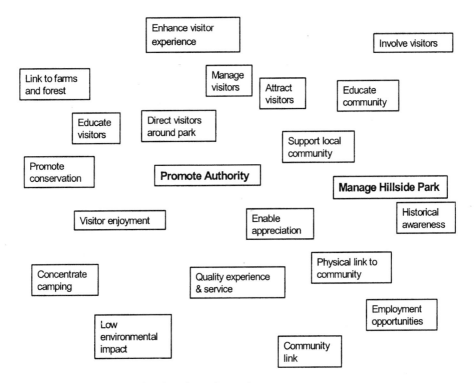

Fig. 3.3 The strategic project functions in random order.

Sorting of functions: high order needs to low order wants

Following the generation of functions undertaken as a brainstorming process, the team are invited to order the functions by putting the highest order need at the top left-hand corner of the paper and the lowest order want at the bottom right. This is illustrated in Fig. 3.4.

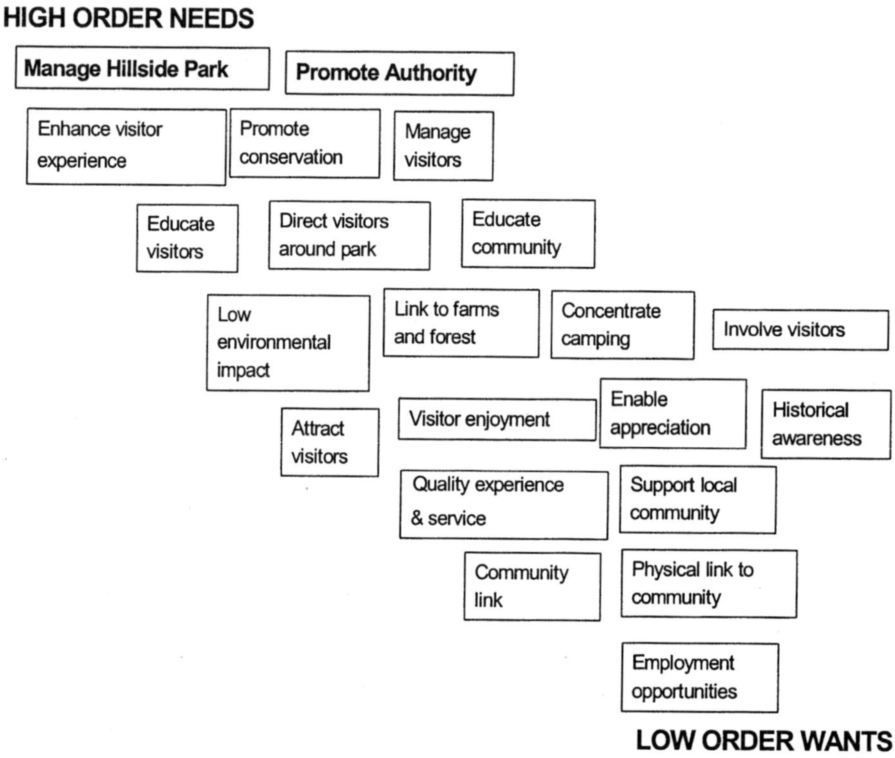

HIGH ORDER NEEDS

Manage Hillside Park	Promote Authority
Enhance visitor experience	Promote conservation	Manage visitors
	Educate visitors	Direct visitors around park	Educate community
		Low environmental impact	Link to farms and forest	Concentrate camping	Involve visitors
			Attract visitors	Visitor enjoyment	Enable appreciation	Historical awareness
				Quality experience & service	Support local community
				Community link	Physical link to community
					Employment opportunities

LOW ORDER WANTS

Fig. 3.4 High order needs to low order wants.

The ordered sticky notes becomes the focus of the function diagram illustrated in Fig. 3.5. The diagram illustrates a client orientated strategic diagram that takes the form of a tree laid horizontally. In a very general sense the logic of the diagram answers the question 'how?' when working from left to right and 'why?' when working from right to left. Highest order needs are placed at the top of the diagram and the lowest order wants at the bottom. A scope line divides the primary function of the project, or the project mission statement, from the functions that may form the basis of brainstorming. At the strategic level, the primary function of the diagram is to understand the 'whole' project in the context of a strategy, programme and projects. In the context of the Hillside Park project, the three elements are:

Function Analysis 63

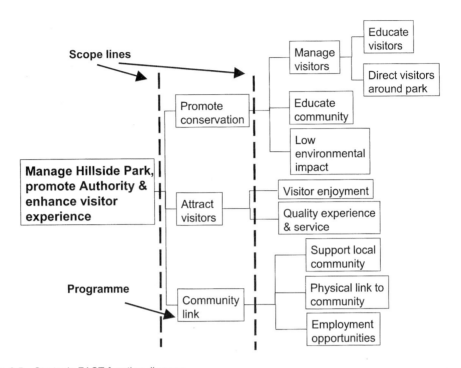

Fig. 3.5 Strategic FAST function diagram.

(1) The strategic mission: manage Hillside Park, promote the authority and enhance the visitor experience.
(2) The three programmes: to promote conservation, to attract visitors and to link with the community.
(3) Projects: in the case of the link with the community this would be to engage the community with regard to authority support and employment opportunities, and to construct a physical link comprising a footpath and bridge.

Function diagramming at the project level

Function diagramming at the project level continues in the same form as above. First however, it would be necessary to consider whether the team that was appropriate for the strategic function diagram is also appropriate for the consideration of the project. The stages in respect of the project are:

- Re-examining the issues and refocusing the team on the problem with regard to the physical link with the town.
- Brainstorming functions in a similar way to that described above. This might give rise to the functions illustrated in Fig. 3.6.

64 Method and Practice

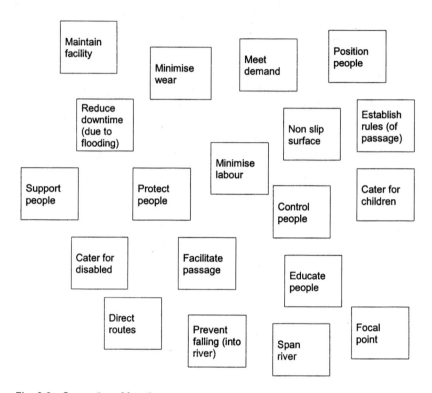

Fig. 3.6 Generation of functions.

- Ordering the functions from high order needs to low order wants in a similar way to that described above, as illustrated in Fig. 3.4.
- Refining of the ordering of functions to construct a matrix as illustrated in Fig. 3.7 that further sub-divides the functions into; strategic needs, strategic wants, tactical needs and tactical wants. Within each quadrant of the matrix the sticky notes are organised by putting the highest order need at the top left-hand corner of the quadrant and the lowest order want at the bottom right. This refinement is especially useful where there are a significant number of both tactical and strategic functions to sort. Various diagrams as illustrated below may now be constructed. In some instances the Task FAST function diagram in Fig. 3.9 is only constructed of the strategic needs and the strategic wants. This technique has the advantage that it speeds up the process of constructing the diagram and further introduces team buy in.

A diagram is constructed from the ordered sticky notes. The type of diagram will depend upon the focus of the study being undertaken. To gain a technical appreciation of the problem a function diagram with a technical bias will be

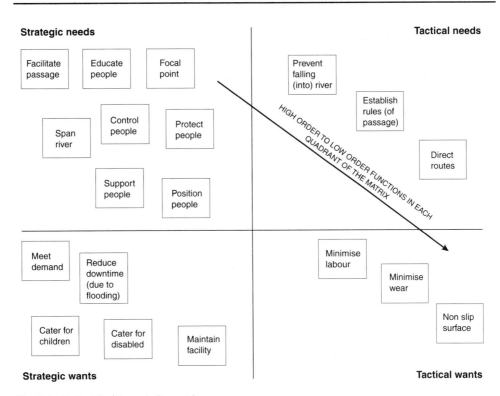

Fig. 3.7 Project function priority matrix.

constructed. If the focus of the study is of a strategic nature, such as a strategic briefing, a client orientated function diagram will be constructed. Figures 3.8 and 3.9 illustrate the two types of diagramming technique for the problems outlined above.

It should be noted from Fig. 3.8 that the prime objective is a technical objective and the brainstorming of ideas following the construction of the diagram will therefore lead to the exploration of technical solutions. The brainstormed solutions, for example a suspension bridge, a simply supported span bridge, stepping stones, etc., will be audited back against the diagram to determine the extent to which the ideas meet the functions.

The diagram is structured in such a way that the prime objective of the project is situated on the left-hand side of the scope line. The prime objective 'support people' is situated immediately to the right of the project objective. Parallel objectives are below the prime objective, in this case 'span river'. Secondary objectives appear to the right of the prime objective and design objectives are situated immediately above. Desired objectives are located at the top right of the diagram above the secondary objectives.

It is the structuring of the diagram that prepares the team for brainstorming. For a client or customer orientated situation the same sticky notes will be ordered in a similar manner to that shown in the strategic diagram Fig. 3.9. It should be noted

66 Method and Practice

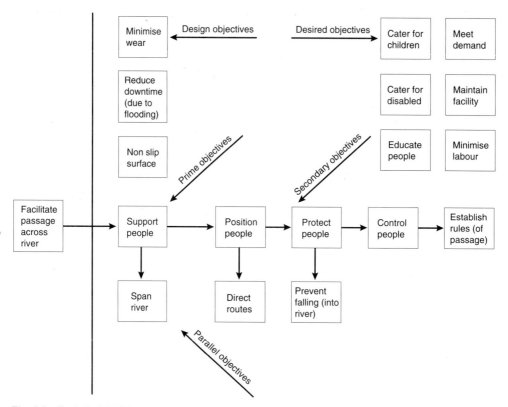

Fig. 3.8 Technical FAST.

that there is a high correlation in terms of the post-it notes used but the emphasis, as judged by the team, has been altered somewhat. In the case of this particular project the final solution might not be a bridge but an old-fashioned cable stayed ferry boat which would use the power of the river acting on the rudder to move it across the river. A full-time ferryman working between two covered shelters, one of which might be an exhibition centre, could operate the ferry.

The final diagraming technique known as critical path FAST (Fig. 3.10) is used in situations where a process is being undertaken. Process in this context is a number of activities in a logical sequence having a beginning and an end.

3.5 Kaufman's FAST diagramming

Kaufman[3] introduces a number of paradigms into the function diagramming process that result in the diagram being read in part from their method of construction. In other words the rules themselves introduce meaning into the diagrams. The annotations to Kaufman's diagramming illustrated in Fig. 3.12 is as follows:

Function Analysis 67

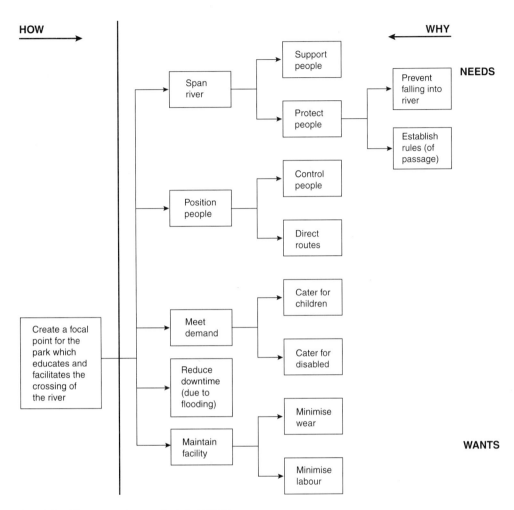

Fig. 3.9 Client or customer orientated FAST.

- *Highest order function*: the objective or output of the basic function.
- *Lowest order function*: a low order function turns on or initiates the subject under review.
- *Basic function*: the mission of the study under review.
- *Concept*: describes the approach to achieve the basic functions.
- *Objectives or specifications*: not functions, but may influence the method selected to achieve the basic functions and satisfy user requirements. (Optional on diagram.)
- *Critical path functions*: any function on the how/why logic path. If it is at the same level as the basic function it is a *major* critical path; otherwise it is an independent or supporting function and a *minor* critical path.

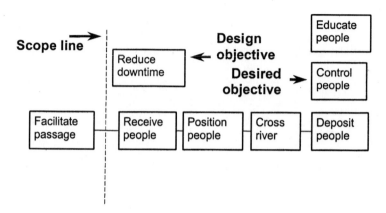

Fig. 3.10 Critical path FAST.

- *Dependent functions*: those dependent on the functions to the left of the basic function answering the how/why logic.
- *Independent (or supporting) functions*: secondary with respect to scope and appearing above the critical path line.
- *Activity*: the method stated to perform a function (or group of functions).

Kaufman also introduces additional rules into the drawing of the diagram such that the importance or hierarchy of the functions can be determined by reading the way in which the diagram is compiled. These rules are illustrated in Fig. 3.11.

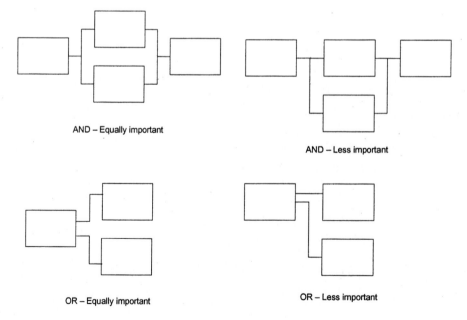

Fig. 3.11 OR and AND diagramming rules.

Function Analysis 69

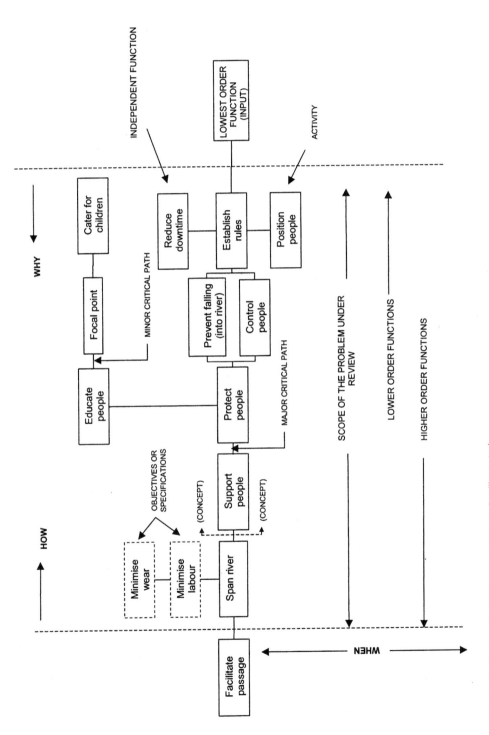

Fig. 3.12 Kaufman's FAST diagramming rules illustrated.

This section has outlined the functional analysis system technique methods described by a number of authors but notably by J.J. Kaufman[4] and by Snodgrass & Kasi[5] in their book *Function Analysis: the Stepping Stones to Good Value*. The functional analysis system technique (FAST) is a useful closing technique that involves the team in a structured exercise to highlight the primary mission of the project. In this respect, as part of the building of the client's value system, the Task FAST diagram reigns supreme.

The section on function diagramming is important because it rehearses in front of the design team all of the factors that need to be taken into account in the design exercise. It is recommended that members of the client organisation outline the daily activities of users and then the client and the design teamwork together to derive the function diagrams. In this way the functional specification leads to a more exact brief including an accurate set of room data sheets.

3.6 Functional space diagramming

Introduction

In section 3.4 the function diagram addresses why a project exists and its main functional attributes. This section, following the decision to build, describes a series of techniques for understanding a structure subdivided into functional space. The form of function diagramming is concerned with the specification of the space to be contained within a structure, considered to be a necessary precursor to the writing of the concept or technical brief. It is necessary to understand that all space within a building performs a function for if space within a building does not perform a function the space is wasted and cost is incurred to no value. For maximum efficiency each space should have the highest degree of usage consistent with its function and therefore it is necessary to ensure that its timetable is reflected within the brief. A further necessary action for the designer is to ensure that circulation space, defined as essential non-functional space, is kept to the minimum consistent with the requirements of organisational efficiency.

The references here to organisational efficiency imply that the space should be configured to maximise the value of the business project expressed in terms of the client's value system and the client's proposed organisational structure. It is vital at this stage to recognise that the very fact that the client is undertaking a project means that the client organisation is about to change. The organisational structure referred to here is the one which will be in place once the change has taken place.

Activity 1: determine users

The first stage in functional space analysis is to identify all of the users of the building. Invariably this will be a longer list than at first anticipated. Figure 3.13 lists the users of a fictitious law court project.

Function Analysis   71

- Judge
- Judge's clerk
- Social worker
- Chairman of children's panel
- Solicitors
- Civil litigants
- Administration staff

- Police
- Custodial accused
- Non-custodial accused
- Jurors
- Witnesses
- Press
- General public

Fig. 3.13 List of users.

Activity 2: flowcharting exercise

Each user from the list in activity 1 is studied in turn and a flow chart prepared of their use of space. This is undertaken by anticipating each activity as part of the user's daily routine within the building where each activity is connected by arrows to the next activity. It is presumed that each activity will require space. Even the activity of entering the building will require an entrance lobby of some sort, and the activity of moving from one space to another indicates circulation space. Figure 3.14 illustrates simplistically the activities of a judge's daily routine.

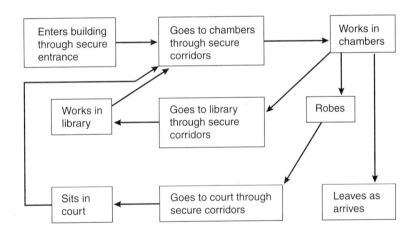

Fig. 3.14 User flow chart.

Activity 3: space specification

Each activity undertaken by the user will require space and that space will have the attributes of size, servicing (heating, lighting, ventilation, acoustics, etc.), quality (normally defined by fittings and furnishings) and finally the technology support required. Activity 3 is a necessary precursor to the eventual compilation of room data sheets. Much of the information that is contained within the space specification will be absorbed into room data sheet specifications. On completion of activity 3 all of the required spaces to be contained within the structure should have been identified and their attributes understood. It is important for this process to remain dynamic as it is easy to get bogged down if spaces are considered one at a time from the flow charts. It is more efficient to list all spaces from the flow chart and then group those spaces that are similar in terms of their attributes.

Activity 4: adjacency matrix

Each space is identified with a distinct name. These names are transferred to the adjacency matrix diagram (Fig. 3.15) that illustrates the adjacency requirement on an index scale of +5 (spaces are required to be adjacent) to −5 (spaces should be designed so that they are not adjacent). In this context adjacency means that there is a physical link between one space and another, normally a door or a short length of corridor. Spaces with an adjacency index of 3 will be within easy reach of one another separated by for instance one flight of stairs or a reasonable length of corridor. Spaces with an adjacency index of 0 give the indication to the designer that the spaces have no adjacency importance one with another and therefore can be anywhere in relation to a the total structure. Spaces with an adjacency index of −5 should be completely separated one from the other in terms of environment, sound and physical linkage. This does not mean that from a geometrical perspective the spaces can't be separated by a single wall. However, it is presumed that one space can not be accessed from the other without travelling through many other spaces. A good example would be two bedrooms in separate apartments within an apartment block. A solid wall separates the bedrooms and to get from one bedroom to the other would necessitate leaving one apartment and entering the one next door. In this situation separation of physical and acoustic environments is highly important.

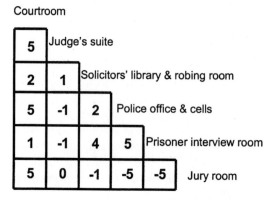

Fig. 3.15 Adjacency matrix.

Function Analysis 73

Activity 5: rationalisation prior to the preparation of room data sheets

A final study is undertaken of all spaces with similar services and environmental attributes. For example, in a study of one building it was determined that the conference centre and the employees' sports and social club both had the same structure and servicing requirements, both were two-storey height spaces and both required high levels of ventilation. The two spaces were therefore placed geometrically together although under the rules of activity 4 both have an adjacency index of −5. The brief highlighted this situation and left the designers to ensure environmental and acoustic separation. In the final design the entrances to the two spaces were entirely remote yet both spaces were able to share a dedicated plant room.

A space usage exercise is carried out to ensure that spaces with similar functional specification are used to the highest degree. For example, if the functional space analysis has highlight spaces of identical functional specification that do not conflict on the organisation timetable the client would need to decide whether these two spaces could indeed be combined. Finally under this section and prior to introducing value engineering it should be stressed that the adjacency matrix and user flow can be a powerful audit tool during a value engineering exercise when analysing current designs.

3.7 Elemental cost planning and elemental cost control

In 1969 the Building Cost Information Service (BCIS) of the Royal Institution of Chartered Surveyors in the UK defined an element as follows:

> 'An element for cost analysis purposes is defined as a component that fulfils a specific function or functions irrespective of its design, specification or construction.'

The introduction to the BCIS document *The Standard Form of Cost Analysis* states that the list of elements is a compromise between this definition and what is considered practical. However, apart from the elements within the subsection building services, the definition works well. This is fortuitous for all those involved in value engineering exercises that have a cost plan in elemental format as, by definition, the building costs are distributed according to element function.

Elemental cost planning is one of a family of techniques based upon parametric modelling. The technique relies upon an extensive database of building costs broken down into elemental costs. In the UK it is normal practice for lump sum, fixed price contracts to be tendered based upon detailed bills of quantities. The bills of quantities are normally arranged in elemental format such that following the selection of the lowest tender an elemental cost analysis of the project is prepared. Quantity surveyors submit the elemental cost analyses to the BCIS. The cost analysis is presented as a list of element costs, expressed as both element costs per square metre of gross floor area and per element unit cost. For example, internal doors will be expressed as £10.50 per square metre of gross floor area and £450.00 each. In this way quantity surveyors share tender data for a wide range of construction projects.

To use the data for cost planning an office building of say 2000 square metres the surveyor uses the rate per square metre to arrive at an approximate cost for internal doors, i.e. £10.50 × 2000 m^2 = £21 000.00. Obviously, the surveyor will need to make a large number of adjustments, for example:

- Inflation in prices between the date of tender of the analysis and the date of tender of the proposed project.
- Difference in prices between the location of the project represented by the analysis and the proposed project.
- Any major differences between the likely specification of the proposed project and the analysis, for example the type and extent of air conditioning, inclusion of car parking and access roads, etc.
- Differences in the market prices due to demand for construction work.
- Differences in risk costs brought about by choice of a particular procurement method.

After these and a large number of other adjustments are made the surveyor will have an elemental cost plan that displays a high degree of cost certainty. This is the point when the cost plan is given to the client for the building and thereafter the budget becomes firm. It should be noted that the cost plan compiled in this way can precede sketch design but rarely does. When sketch drawings are available, using the example above, the surveyor can count the actual number of doors, say 40 nr. The cost for the internal doors element can then be refined as £450.00 × 40 = £18 000.00. The overestimate of £3000 will be added to the contingency to pay for those elements that may have been underestimated.

Two important points to note here are:

- First, while value engineering will address the functions of all elements, cost planning generally only triggers action in the event of an overspend. In the example of internal doors, value engineering will seek to identify unnecessary cost in the function of the element whereas the action of cost planning will only trigger action when the element in question is judged to be overspent by reference to the cost plan.
- Second, the sketch design will assume the client's value criteria. If these have not previously been discovered the designer will assume them. The debate on the alignment of project and client value systems is in Chapter 8.

3.8 Element function analysis

While the concept of elemental analysis was derived for the cost planning function described above, two ingredients are essential to the undertaking of elemental function analysis. These ingredients are:

- A data base of costs which may be used for benchmarking projects.
- A common understanding of the costs which are contained within a particular element. For example, the BCIS definitions will include the costs of forming the opening for a window in with the cost of the window element.

Element function analysis comprises the stages listed below.

Stage 1: identify the cost dominant elements

In practice it is not realistic to value engineer all 34 elements represented by the BCIS cost analysis. Therefore some method must be derived for determining which elements at first sight appear to be offering poor value for money by being either unreasonably expensive or indeed unreasonably inexpensive. The production of a histogram of element costs is a useful way of presenting data in order to make an informed decision. Often the attention of the value engineering team is directed towards those elements containing the largest proportion of the total cost and those elements that appear to be uncharacteristically expensive. The latter will be determined by building type.

For example, the element roof is expected to have a proportionately high cost in a single-storey primary school whereas an acute hospital will tend to have proportionately high costs for the element services. These elements may be worthy of investigation purely on the basis of their high proportion of total cost but may also serve as a benchmark to trigger investigations in other elements, e.g. if the external walls were significantly the highest cost element in a single-storey school then the relative position of roof to external wall in the cost plan histogram might trigger an investigation into the external wall element. Notwithstanding this, it is often the case that an element will appear to be offering reasonable value for money but can still be value engineered without loss of function.

Stage 2: list all the functions of the selected element

As elements are defined as being components of construction that fulfil a specific function or functions irrespective of its design, specification or construction it is logical to deduce that a definitive list of functions can be derived for each element in the BCIS list. See Table A1.2, page 283.

For example, an internal wall will have one or more of the following functions irrespective of the project context:

- Support load
- Divide space
- Separate environments
- Attenuate noise
- Transmit light
- Secure space
- Support fittings
- Facilitate finishing
- Restrict fire spread
- Demonstrate hierarchy
- Minimise distraction

Stage 3: select functions for project context

In this next stage of element function analysis the list is reviewed and functions deleted which are not relevant to the project situation.

An internal partition is a particular case of an element that may have a number of different functions within the same building. Therefore it would be necessary for the team to undertake a study of internal partition type before proceeding further. For example, such a study of a university department building may reveal the following partition types:

- Division between lecture rooms
- Division between laboratories
- Division between storerooms
- Division between lecture rooms and corridor
- Division between offices
- Division surrounding computer room

The above illustrates a number of partition types that display differing functional characteristics. For example, a division between two lecture rooms is required to 'attenuate noise' to an absolute minimum while a division between stores needs only to 'divide space' and perhaps 'support fittings'. In each design situation the partition as designed is compared to functional need. The review may highlight the functional properties of the partition as inadequate and/or it may be wasteful.

To take the functions of a partition between lecture rooms as an example of function selection, the process would be first to delete those functions that do not apply in this situation (see Fig. 3.16).

Function	Reason for deletion or retention
~~Support load~~	Framed building
Divide space	Required function
~~Separate environments~~	Heat, vent, etc. – same requirement either side
Attenuate noise	A primary requirement
~~Transmit light~~	Not required
Secure space	Lecture room contains IT equipment
Support fittings	Boards, screen, display panels, etc.
Facilitate finishing	Hard surface finish, easy to clean
Restrict fire-spread	Required function
~~Demonstrate hierarchy~~	Not required
Minimise distraction	No visual or other sensory transmission is permitted between lecture rooms

Fig. 3.16 Element clusters.

Stage 4: brainstorming solutions

The brainstorming exercise will be undertaken on an element by element basis following the complete analysis of the elements. In the partition example ideas are generated to meet the retained functions. The rules relating to brainstorming are described in Chapter 4.

Stage 5: evaluation and development

During the evaluation phase the large number of ideas generated through brainstorming are reduced through a logical process of option reduction. In a situation similar to the partition example above, where the number of technical solutions are limited, the use of a weighting and scoring matrix is useful (see Toolbox). The development stage takes the highest scoring or most promising technical solution for further technical development. The preferred solution must pass the test that it is technically feasible, economically viable, functionally suitable and acceptable to the client.

3.9 Element function debated

Figure 3.17 highlights the characteristic feature of all buildings which is that key functions are allied to groups of elements. For example the substructure, frame and upper floors all support and transfer load. On the other hand the internal walls, internal doors and stairs, the wall, floor and ceiling finishes, the fittings and furnishings, the plumbing and the sanitaryware directly impact the client's organisation. The client should be able to work with the building and its arrangement, not against it. Therefore it could be argued that a greater emphasis should be made on designing these elements to achieve an accurate fit with the client's organisation rather than the structural elements. Further, it is important that those elements that directly interface with the client's business should reflect the value system of the client.

Figure 3.17 is useful in a workshop situation particularly where it has been decided to cluster members of the supply chain for smaller, shorter, more focused value engineering workshops. However, Fig. 3.17 is only an aid and account should be taken of, for example:

- The structural cluster's responsibility to understand the client's use of the building in respect of the specific live loading, which is related to the client's business.
- The fact that external and internal walls may be load bearing.
- Building regulations require a minimum number of sanitary fittings, fire protection and fire escapes, insulation, day lighting, dry risers, hose reels, fire alarms and detection, external fire hydrants, fire access roads, etc.
- Although internal environment maintains function there is an extra requirement that results from the client's use of the building, primarily impacting air conditioning and specific electrical loads.

3.10 Conclusion

This chapter is an overview of function analysis specifically applied to construction. In this context the view has been taken that there are three distinct applications: strategic function analysis, function space analysis and element function analysis. There are a number of texts and papers that refer to function analysis as an abstract technique, where examples are given they are generally given as examples of the function analysis of fully costed elements. This chapter is a synthesis of that literature together with data derived from action research studies. The developments described in Chapter 2 are drawn together with the data in this chapter to form a holistic approach to the value management service, which is developed and discussed in Chapter 5.

It has already been stated that the three things which make value management and value engineering different from other management techniques are the three factors of making explicit the clients value system, the application of function analysis and the use of teams. The next chapter examines the theory and practice of teams, team dynamics and facilitation.

Function Analysis 79

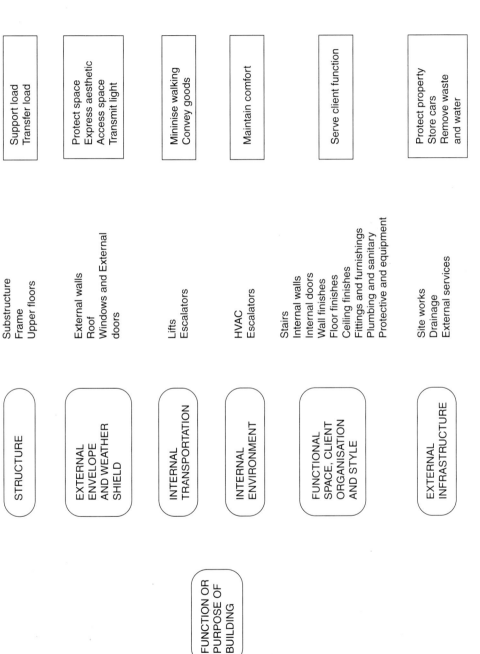

Fig. 3.17 Functions allied to building elements.

3.11 References

1. Morris, P. W. G. & Hough, G. H. (1987) *The Anatomy of Major Projects: A Study of the Reality of Project Management.* Wiley, Chichester.
2. Bytheway, C. W. (1965) Basic function determination technique. *Proceedings of the Fifth National Meeting of the Society of American Value Engineers,* **2**, 21–3.
3, 4. Kaufman, J. J. (1990) *Value Engineering for the Practitioner.* North Carolina State University, Chapel Hill, NC.
5. Snodgrass, T. J. & Kasi, M. (1986) *Function Analysis: the Stepping Stones to Good Value.* University of Wisconsin, Madison, Wisconsin.

4 Teams, Team Dynamics and Facilitation

4.1 Introduction

Value management and value engineering are facilitated team activities. This chapter summarises current thinking in respect of groups and teams and outlines characteristics and methods of the facilitation of value management and value engineering teams. In this context teams tend to be a collection of individuals undertaking a specific stakeholder, project management, constructor or professional role within the project. The choice of the individuals for the team is more related to job function that it is to individual traits – an important factor that should be born in mind in the study of this chapter.

Like the previous chapter on function analysis, this chapter is a synthesis of literature and the conclusions from action research activity. The characteristic of value management and value engineering as facilitated team activities is one part of the formula that makes value management and value engineering unique as a management technique.

4.2 Groups

Schein[1] defines a group in psychological terms as any number of people who interact with one another, are psychologically aware of one another and perceive themselves to be a group. Hellriegel et al.[2] define a group as comprising people with shared goals who often communicate with one another over a period of time and are few enough that each individual can communicate with all the others person to person. Cook et al.[3] define a group as two or more people who regularly interact with and influence one another over a period of time, perceive themselves as a distinct entity distinguishable from others with shared common values, and are striving for common objectives. While these definitions are fairly wide they are useful because they highlight the interaction and awareness within the group and distinguish them from other gatherings of people, for example a bus queue or guests in a hotel. They all also imply that individuals within a group will sacrifice individualism for collectivism.

Groups can be divided into formal groups that are especially created to perform tasks, and informal groups which tend to be more socially motivated. Formal groups can be divided into three types:

- An equalisation group formed to resolve some type of conflict usually through negotiation and compromise. The group initially is formed of individuals with contrary opinions but with a willingness to seek equalisation or balance. The group goal is a solution acceptable to the group as a whole.
- A cooperative group where members rely on each other to perform tasks which result in an individual gain at the end of the project; for example, a group of students attending classes and working on group assignments with an individual assessment or examination. A cooperative benchmarking group would also fall into this category. The group goal is the success of the individual.
- A project focused interacting group where individual members rely on the others to the extent that the group goal cannot be achieved unless each individual has undertaken their task. In this type of group individuals bring to the table complementary skills to achieve the group goal which is the success of the project.

Formal groups can also be considered as:

(1) *Permanent* formal groups that exist with the knowledge that they are unlikely to be disbanded, for example a standing committee or an accounts department. Individual group membership may change over time but the committee or department continues in existence.
(2) *Temporary* formal groups which exist knowing that at some time in the future they will cease to exist as a group. Temporary formal groups have a life cycle from formation through to cessation and they can exist for long periods of time. The difference between temporary and permanent groups is not one of longevity of existence but of certainty of the end.

Informal groups emerge to satisfy personal needs for support, friendship and recreation. Groups may be defined as interest groups, friendship groups and reference groups where the latter is one which an individual joins for the purpose of forming opinions, making decisions or determining how to act. Informal groups in a work related context might emerge out of formal group activities and satisfy social as opposed to organisational needs. In this context informal groups may emerge as a function of the probability of interacting that is often related to physical proximity. Informal groups can subvert on-going formal activities if they form an identity of their own which is counter to that of the organisation. In large organisations informal groups can be broken down into the following types[4]:

- Horizontal cliques, where group members are of near equal organisational status and may work in close proximity.

- Vertical cliques, where group members may be from different organisational status levels within the same department or team and know each other either from non-work related activities or because they need to achieve particular goals relevant to common interests.
- Mixed or random cliques, whose members are from different organisational status levels, departments and locations. This type of group may emerge to serve common interests or fulfil objectives that the formal organisation is unable to meet.

Emphasised here is the fact that neither value management nor value engineering are group activities but are team activities, a concept which is to be explained in the next section.

4.3 Teams

A team is defined by Hellriegel et al.[5] and Cook et al.[6] as a type of group with complementary skills, competencies and knowledge, who are committed to a common purpose, set of performance goals and approach for which they will hold themselves mutually accountable. A team engages in collective work and through coordinated joint effort produces results that are more than the sum of the individual efforts. A team is characterised by a shared commitment by all the members to their collective performance and this will influence how they work together to accomplish the team's goals. Empowerment is therefore an important feature of team working and the limits of that empowerment need to be fully understood by each member of the team. Value management and value engineering are team activities with defined goals and a structure defined by the facilitator. Empowerment is generally controlled through the process although as discussed in Chapter 5 there are times when the confident facilitator will empower the team to achieve greater performance. In sum, a good team is characterised by[7]:

- A clear sense of itself as a group.
- Positive interaction with outsiders.
- The cultivation of positive assumptions and beliefs.
- Communicating clearly.

Construction project teams are temporary formal teams that bring together the knowledge and skills of people from various professional and business backgrounds to identify tasks and solve problems in the realisation of the construction project. In developing potential solutions they are empowered to take action within defined limits. As Male[8] stated, project teams in the construction industry can comprise people either from within the same organisation (intra-organisational project teams) or from different organisations (inter-organisational project teams), a design team being one example. Value management teams may comprise both. Project teams in the construction industry have also been termed 'temporary multiple organisations'[9] or project coalitions[10], the former label denoting the fact that project teams comprise representatives from different organisations brought together to achieve a particular task while the latter denotes that power structures will also exist within project teams between different organisational representatives.

Team dynamics

Team dynamics or processes are affected by the following factors[11]:

(1) Team size.
(2) Membership roles and composition.
(3) Group norms concerned with emergent implicit or explicit codes of behaviour.
(4) Group goals: the task to be performed or objectives attained.
(5) Group coherence.
(6) Leadership, both formal and informal.
(7) External environment: those factors outside of the group that impact its operations but over which it may have limited control.

All of the above is relevant to the dynamics of the value management and value engineering team and has to be considered in detail before engaging with the team at a workshop.

Team size

The size of an effective team can range from two members to an upper limit of about 16 members although 12 members is probably the largest size that allows each member to interact easily face to face[12]. Optimally, the team size is between six and ten people[13]. It should be noted, however, that members of small teams of two to seven interact differently from members of large teams of 13 to 16. In some circumstances it is necessary to sub-divide a large team to form a sub-team of between five and seven members to consider specific matters in greater detail.

In small teams there is a low tolerance of team members to the leader's direction but demands on the leader in terms of direction and control are low. In large teams there are higher demands on the leader but team members will be more tolerant of the leader's direction. Similarly, factors such as team member inhibition are low in small teams and high in large teams. Larger teams require more formal rules and procedures and will typically take more time in reaching decisions. It should be noted that particularly value management teams tend to be large (18–20) and very rarely are below ten. This fact becomes obvious if consideration is given to the identity of members at, say, a charette. A charette would comprise key members from the client organisation, say six to eight in total, plus facilities manager, architect, surveyors, structural, mechanical and electrical engineers, project manager, construction manager, key members from the contractor and the contractor's supply chain, say four or five in total, bringing the team to a minimum of 19.

In contrast, dependent on task a value engineering study can be small, often not exceeding six to eight members.

Membership roles and composition

The membership and composition of the team will influence behaviour, dynamics and outcomes, which are a function of the similarities and differences within the team membership. In this context each team member has a set of expectations about how they as an individual will behave and also holds expectations, beliefs and assumptions about how other people will behave. These two sets of impressions combine to give an individual a situational view of the world, termed a role[14]. Hunt[15] indicates that studies of teams have demonstrated the above in two behaviour patterns, which he refers to as task orientated and maintenance orientated behaviour. Hellriegel et al.[16] reinforce this by stating that all team members will perform task orientated, relations orientated and self-orientated activities.

- Task orientated activities relate to the project, for example initiating new ideas, seeking information, giving information, coordinating and evaluating.
- Relations orientated activities relate to encouraging, harmonising, consensus seeking, conflict resolving, integrative processes.
- Self-orientated activities are self-centred and will make themselves apparent through blocking of progress, seeking recognition, domination, etc.

Maintenance behaviour is therefore comprised of relations orientated or self-orientated behaviour. Hunt suggests that contributors to the team will, over time, sort out task and maintenance behaviour amongst themselves in order to enable the task to be undertaken. An observation by Male, in numerous simulated task centred, multi-disciplinary team situations, also suggests that personal problem solving styles, professional background and personality interact in a complex way in problem solving situations. These observations suggest that where there is a 'fit' between team dynamics and these variables the team is more concerned with task performance. However, if there is a degree of imbalance between these variables the team will spend more time on managing itself rather than concentrating on the task. If there is a severe imbalance, usually in the area of preferred problem solving style and personality rather than professional discipline, the team can 'lock' into managing team processes to the almost total exclusion of task performance.

Belbin[17,18] defines a team role as 'a tendency to behave, contribute and interrelate with others in distinctive ways'. The importance lies in the characteristic behaviour of one person in relation to another or in relation to the progress being made by the whole team. Belbin has developed a psychometric test whereby an individual's underlying personality traits and behaviour can be determined, allowing team matching. The nine personality traits are:

- Action orientated
    - Shaper – challenging, dynamic, thrives on pressure, overcomes obstacles.
    - Implementer – reliable, disciplined, efficient, turns ideas into actions.
    - Completer finisher – conscientious, anxious, seeks errors and delivers on time.
- People orientated
    - Coordinator – confident, seeks goals, delegates and promotes decision making.
    - Team worker – cooperative, diplomatic and averts friction.
    - Resource investigator – extrovert, enthusiastic, communicative, seeks opportunities and contacts.
- Cerebral role
    - Plant – creative, imaginative and unorthodox in solving difficult problems.
    - Monitor – strategic, seeks all options and evaluates accurately.
    - Specialist – single-minded, dedicated, providing knowledge and skills within a narrow focus.

Belbin's basic premise is that teams should be balanced or at least aware that if, for example, there is no completer finisher in the team then the probability of completion on time is remote.

As stated above, the choice of particular team members for value management and value engineering studies is more determined by job function than by individual trait. This is a danger of which the facilitator should be aware.

4.4 Team norms

Team norms are rules and patterns of behaviour that are accepted and expected by members of the team. These emerge from team interactions and the shared expectations of contributors. They will either overtly or covertly control member behaviour in a manner that members believe to be necessary to help them reach their goals. Team norms will take some time to emerge and therefore the longer the team is operating together the more likely that strong norms will develop. Hunt[19] contends that the degree of conformity to team norms will depend on:

- A person's desire for the team to accept his or her membership – acceptance.
- A person's desire to avoid displeasure, punishment or isolation from the team – pleasure.
- A person's belief that team norms are a reflection of personal views – congruence.
- A person's ability to handle the doubt that they may not be able to stand alone – isolation.
- A person's belief in team goals – agreement.
- The development of team norms also operates in the context of the structuring within a team.

These issues are important for the new entrant to an established team, as might be the case with an established client team. The value management facilitator is in danger of being the only one who is not a team member.

4.5 Team coherence

Team cohesiveness reflects the strength of the members' desire to remain in the team and their commitment to it. This will depend in part on the frequency of team meetings, the importance of attaining objectives through task performance, and the degree of compatibility between team goals and individual member's goals. Where team members have a strong desire to remain in the team and personally accept its goals this leads to highly cohesive and possibly powerful teams. In a value management context the team will meet so infrequently that this is not an issue. However, an established team undertaking value engineering exercises will find this a factor to be aware of.

4.6 Leadership

Informal leadership tends to grow over time and usually reflects a unique ability in an individual to help the team reach its goals. Formal leaders are those that are enforced upon the team or are appointed to lead the team. In formal work teams, hierarchy, authority and managerial style affect team structuring[20]. There may be a formal leader appointed to manage the team but sometimes an informal leader(s) may also emerge in addition to the formal leader. Occasionally, depending on the balance between the perceived expertise of the formal leader and those who perform task centred and maintenance behaviours, the formal and informal leaders may clash. Team norms will have a very strong guiding influence on who is ultimately recognised as the leader of the team. This will be discussed in more detail with reference to value management teams in the section on facilitation and also in Chapter 5. In a normal value management situation the facilitator becomes the formal leader. The issue is then how much maintenance behaviour does the facilitator allow the team recognising the problems which may arise with an informal leader.

4.7 Team development

Bringing together the views of Adair[21], Heiriegel et al.[22] and Hunt[23], the development of a team is represented in the following stages:

- *Forming*: Getting to know each other. Task orientated behaviour is attempting to grapple with the requirements of task performance. Maintenance behaviours are concerned with socialisation issues, overcoming anxieties and working out power relationships.
- *Storming*: Power struggles emerge. Competition and conflicts between contributors emerge and the formal leader may be challenged. The management of conflict rather than its suppression becomes essential. Polarisation of ideas can occur as can psychological withdrawal from the team. The feasibility of the task is questioned.
- *Norming*: Conflict is reduced and power is distributed. Rules of behaviour emerge implicitly or explicitly and the team identifies a sense of common responsibility for task performance. Task orientated behaviour ensues in the form of the free sharing of information and the acceptance that opinions may differ, and the team searches for compromise on tasks and objectives. Maintenance behaviours are directed towards establishing cohesion and support.
- *Performing*: Task performance. Contributors' roles are clearly identified, understood and accepted. A frank exchange of facts, opinions and preferences occurs. Trust is established and problem solving is free flowing; decision making occurs and the team experiences a high degree of cohesion. The team has worked out a structure to achieve the task or objective.
- *Adjourning*: The task is complete and the team disbands. There may be signs of reluctance or regret.

The issue here for the value management facilitator is whether the forming, storming and norming situations are allowed to develop. There is an argument which states that to only perform and adjourn is highly efficient though it may detract from good team building.

4.8 Team think

A phenomenon called 'team think', a term coined by Janis[24] and discussed in more detail by Janis & Mann[25], can develop as the team development process proceeds. It tends to occur in highly cohesive, conforming teams and depends upon the balance of expertise, power distribution, problem solving styles and leadership behaviour coupled with the degree of team isolation from outside influences.

The characteristics of team think are:

- Task orientated behaviour is terminated.
- Disengagement from maintenance behaviour.
- The illusion of invulnerability leading to overoptimism and extreme risk taking.
- Team rationalisation without re-validating assumptions prior to taking action.
- An absolute belief in the moral integrity of the team;.
- Stereotyping of outsiders.
- Uncompromising pressure to conform to team norms.
- Uncritical thinking and self-censorship.
- The illusion of unanimity.

Hunt[26], based on his own observations, adds an additional three characteristics:

- A belief that all contributors have expressed their views and that the outcome results from a consensus of divergent views.
- A concern for any answer regardless of its merits.
- A failure to identify expertise among contributors.

The lesson for the value management facilitator is to be aware of the introduction of newcomers into highly cohesive teams. The introduction of for example, the contractor or specialists into teams which have been working together for months or even years can be a problem.

4.9 Selecting team members

Members of value management teams should be chosen for their ability to contribute information and enable or undertake decision taking. The ACID test is a useful aid to selecting team members.

- **A**uthorise: include those who have the authority to take decisions during the workshop process.
- **C**onsult: include those who have to be consulted during the workshop process and without whose consultation the workshop would be suspended.
- **I**nform: exclude those who merely have to be informed of the outcome of the workshop.
- **D**o: include those who have to translate the outcomes of the workshop into action.

In selecting members of the team, multiple membership of one department or organisation should be avoided where a cross section of views is being sought. For example, it may be unhelpful to have three architects plus one representative of each of the other professions of the design team. Similarly three members of the accountancy department might dominate a team with one member of each of the other departments of the client's organisation.

4.10 Facilitation

A facilitator is a formal leader whose presence can frustrate the normal development of a team such that the forming, storming and norming stages described above can be negated. In the context of a value management exercise, therefore, a skilled facilitator can efficiently manage a temporary team so that maintenance behaviour is minimised and task behaviour maximised. It is important, however, from a project management perspective that the leadership roles of the value manager and project manager are recognised and respected. Having said this, it is highly likely that members at a value management study may not have met before and certainly in the case of client members are not likely to meet the design and construction team formally again.

It is appropriate to review the practical application of value management from the perspective of the facilitation of the process for a number of reasons:

- Value management is primarily a facilitated process.
- Where a value management service is sought it is often the case that a value management facilitator will be contacted.
- The value management facilitator is often the only person in a team who is knowledgeable and skilled in the value management process.
- Value management has application at certain strategic points in the development of a project and therefore the facilitator tends to be engaged only for the number of hours required to undertake the value management exercise.

4.11 Facilitation defined

Facilitation involves the controlling and leading of a team through a process using analytical, arbitration, guiding and influencing skills. Facilitation is distinguishable from chairmanship in that the facilitator is not a member of the team, contributes nothing more than the facilitating skills, and has no vote and certainly no casting vote. Their work is not recorded in the final report of the value management study. The facilitation work therefore becomes invisible once the value management study is complete. In any leadership or decision making role it is very dangerous to set oneself up as the arbiter of normality and this is never more true that in the role of the facilitator.

4.12 Identity and role of the facilitator

There are no set rules on what type of person makes an ideal facilitator[27]. For example, introverts appear to make good facilitators as well as extroverts. However, a common feature of most facilitators is that they have self-selected themselves for that role and gain enjoyment from it. There are, however, some practical aspects regarding the sourcing of facilitators and some generally accepted rules.

Internal to the team

Internal to the team. It is generally accepted that a member of a working team does not make a good facilitator as it is not practical to both undertake a team role and facilitate at the same time. For example, were a project manager to facilitate the team being managed then independence would be lost and the exercise could become fraught with hidden agenda.

External to the team but internal to the practice or company

Being external to the team is the preferred position of a facilitator and therefore there is no reason why the facilitator may not be internal to the practice or company. However, in this situation one primary rule will have to be observed: while acting as a facilitator a company employee assumes a position, in the eyes of the team, equivalent to the chief executive or highest ranking officer. This is important as a facilitator often senses that the team sees the necessity to explore an issue in detail which may be uncomfortable for one or more members of the team. Using a military analogy, if the facilitator is a lieutenant and the uncomfortable member of the team is a major, the major may pull rank to ensure that the particular issue is not explored. Organisations with considerable value management activities and in-house facilitators have positioned the facilitation organisation as responsible to the board of directors or for example in one local authority, to the treasurer's department. This may ensure cooperation as cooperation may be a prerequisite to obtaining funding for a project.

The same rules do not apply to the same extent with facilitators from the same consultancy organisation. For example, as stated above, a project manager would not make a good facilitator on the project for which he is responsible. However, a consultant project manager may be a perfectly acceptable facilitator for a value management exercise on a project managed by a colleague from the same consultancy organisation. The key here is the measure of perceived externality and therefore independence.

External to the team, practice, company or organisation

The truly external facilitator, the consultant facilitator, overcomes all of the issues discussed above since this type of facilitator has no position within a particular organisation and has no 'axe to grind'. The consultant facilitator may be the preferred option for an organisation that does not undertake sufficient value management studies to make an in-house facilitation organisation worthwhile. Additionally the external facilitator's independence will also be an advantage in complex projects with difficult problems to solve and with different factions to be brought together. The external facilitator will bring with them a broad experience of facilitating studies in other parts of the industry that could help with the problem solving required. There should be no question in any participant's mind as to the facilitator's independence.

The facilitator's role

The role of the facilitator involves planning the workshop process and the physical environment, engendering consensus among members of the team within the workshop and recording the result. The key facilitation roles are therefore:

- Ensuring an efficient physical workshop environment.
    - ☐ Select an appropriate room with such characteristics as clear walls for displaying drawings and key information resulting from the workshop and sufficient space to allow the team to be at tables arranged in a horseshoe formation so that all members of the team have eye contact and a clear view of the facilitator. This may be required to change during group working sessions; even better is the use of breakout rooms. The room should ideally be away from any member of the team's normal place of work to ensure an interruption free workshop. Obviously the usual toilet location and fire escape information will be required.
    - ☐ The equipment that is required for the room is standard office equipment including repositional sticky notes, pens, card, etc., with large areas for writing on. The use of flip charts and/or overhead projectors is common with other electronic forms of visual presentation via notebook computers and projectors gaining popularity. However, as with white boards, this information tends to disappear whereas if the record is made on flip charts it can be displayed for the duration of the workshop.
- Setting up the team to maximise an effective outcome.
    - ☐ Assist the client in the choice of team members and ensuring the presence of key decision takers appropriate to the stage in the workshop.

- ☐ Visit key members of the team prior to the workshop to interview the participants and obtain their views on specific issues relating to the project and the study and to promote the value management process, thereby identifying any resistance to the process or any hidden agenda. This will also help in the preparation of the agenda for the workshop.
- ☐ Consider the way in which any hidden agenda will be dealt with in the workshop.
- ☐ Agree the dress code and behaviour code for the workshop, e.g. first names only to help break down hierarchical title barriers. It is important for people to feel at ease and whether this means casual or formal or a mix of dress should not matter.
■ Spending time understanding the project and developing a structured agenda. It may become necessary to depart from the agenda during the team exercise or to spend longer than anticipated on some activities. It may be advisable, therefore, to circumvent queries as to progress by not putting times against activities on the agenda circulated to team members.
■ Understanding the criteria for a successful workshop.
■ Informing all team members of the purpose of the workshop, preferably prior to the workshop. Some facilitators like to establish a facilitation contract with the team but this assumes a fairly high level of VM knowledge within the team.
■ Sending to the workshop participants a few days in advance of the study the agenda, workshop location, details and purpose together with any information that they should bring with them. The date for the study will have been agreed with participants some time before.
■ Managing the process of the workshop.
- ☐ Consider some ice breaking activity at the start of the workshop; introducing your neighbour to the team is an example.
- ☐ Team members should understand that they are attending because they have something to offer not just because they represent departmental interests.
- ☐ Recognise the attributes of team members as individuals and their contribution, restricting irrelevant discussion and story telling and encouraging observations from non-contributors.
- ☐ Ensure that all team members participate in the discussion and encouraging an atmosphere of frankness and openness by, for example, breaking into small groups and then combining, asking questions of those not contributing, asking someone to summarise a previous discussion, watching the body language.
- ☐ Do not avoid issues or difficult questions; they are likely to be important to the outcome of the workshop.
- ☐ Notwithstanding the above point, understand the different stakeholder positions particularly in respect of organisational hierarchy or financial dependence within the team.
- ☐ Understand the differences between facilitating small teams and big

teams. Big teams may require breaking down into smaller groups and this may lead to a loss of facilitator control.
- ☐ Plan the team to include all skills and query with the client where a particular skill is missing.
- ☐ Consider the voting position. When consensus is difficult the alternative is to ask for a vote. However, the facilitator should be aware that not all members of the team have voting rights. For example, for discovering the client's value criteria it would be inappropriate for the client's consultants to vote.
- ☐ Sense interpersonal relationships within the team and avoid emotive argument or point scoring.
- ☐ Sense the climate of the workshop. For example, do not be afraid of silence.
- ☐ Deal with any hidden agenda amongst workshop participants.
- ☐ Provide direction and a sense of common purpose.
- ☐ Question, feigning ignorance, to ensure full team understanding.
- ☐ Summarise after each section to maintain momentum and avoid revisiting earlier discussions unnecessarily.
- ☐ Ensure team members express themselves simply, discouraging TLAs (three letter acronyms) and inappropriate technical terms.
- ☐ Achieve consensus and avoid overt voting.
- ☐ Synthesise and integrate information during the workshop.
- ☐ Ensure everyone understands agreed facts, for example by stating that everything written down in front of the team on a flip chart, board or by electronic display is a record of what is agreed. Ensure the accuracy of the flip chart record by listening, questioning for clarification, paraphrasing and summarising at regular intervals. If anyone disagrees with what is written down it must be deleted and discussed until consensus is achieved.
- ☐ Maintain good eye contact.
- ☐ Maintain momentum.
- ☐ Use first names: name plates required.
- ☐ Facilitate, don't perform.
- ☐ Guide, don't lead.
- ☐ Consider whether to employ a recorder. A recorder taking notes of the workshop proceedings and entering the flip chart information direct into a computer can significantly reduce the effort required in producing the workshop report.
- ☐ Attend to meeting logistics, i.e. a room with plenty of workspace, breakout rooms, flip charts, cards, sticky notes, pens, scales, sticky dots, scissors, masking tape, blue tack, etc.
- ☐ Ensure that comfort breaks are taken and be aware that a comfort break may release a tension point and allow consideration of how best to proceed.
- ☐ Get the balance right (Fig. 4.1).

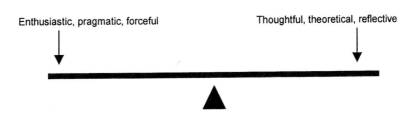

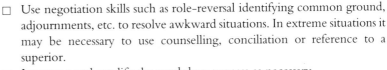

Fig. 4.1 Workshop balance.

- ☐ Use negotiation skills such as role-reversal identifying common ground, adjournments, etc. to resolve awkward situations. In extreme situations it may be necessary to use counselling, conciliation or reference to a superior.
- ☐ Intervene and modify the workshop process as necessary.
- ☐ Communicate verbally and in writing to produce the value management report.
- ☐ Ensure that the team buy in to the workshop rules, for example:
  No mobile phones.
  If you leave, don't come back.
  Anything displayed on the wall is agreed by the team – if you don't agree with what has been written on the flip chart, say so.
  A member of the team – not the facilitator – will do the presentation.
  All participants are equal.
  One person talks at a time.
  Every idea or comment is valid.
  Only valid stories permitted, and then only if concise.
  No judging during brainstorming.
  What is said here stays here – confidentiality.
  The outcomes are the group's.

From the above it can be deduced that a facilitator requires skills in team management, organisation and recording. Additionally the facilitator will need skills associated with assessing the mood of the team. These skills include reading body language in corroboration of what is being said at the time and in what manner. Tension within the group is not uncommon but if allowed to ferment can become destructive.

4.13 Facilitation styles

The benchmarking study[28] highlighted three distinct facilitation styles operated under the banner of VM practice. These styles were defined as:

- ■ Facilitation style 1: facilitate a workshop only.
- ■ Facilitation style 2: facilitate a workshop with some advanced preparation.

- Facilitation style 3: facilitate workshop with full preparation prior to the workshop.

Each style will be discussed in turn.

Facilitation style 1: facilitation only

This style of facilitation assumes no preparation, a reliance on the client to choose the workshop team and a profession by the facilitator of no knowledge of the subject to be discussed. In this situation the opening technique is likely to involve inviting issues from the team and/or inviting presentations from those team members who believe that they have key information.

Following the discussion of issues the facilitator would:

- Summarise the information and gain team consensus on the problems to be addressed.
- Configure the problems in a form conducive to the generation of innovative ideas.
- Direct an ideas generation session.
- Assist team members in the selection of ideas for development.
- Direct team members in the development of the selected ideas.
- Conduct a presentation session following which the best ideas are taken forward for further development.
- Record the workshop proceedings and write a report.

In this style of facilitation, the facilitator is relying totally on facilitation skills for a meaningful workshop. The absence of a key team member and/or vital information from the workshop may result in the full potential of the workshop not being realised. Also, the lack of sufficient project knowledge can result in a lack of focus and an inadequate agenda.

Facilitation style 2: facilitation with preparation

In this situation the facilitator prepares for the workshop by holding a pre-workshop meeting. This allows some influence on the selection of the team for the workshop in that any missing key members can be identified. The prime objective of the pre-workshop meeting is to ensure that all information required by the workshop will be available and that the main issues will have been identified. Having held a pre-workshop meeting the facilitator will conduct the workshop in a similar manner as above but with the confidence that; the main issues have been identified and therefore the agenda for the workshop is appropriate, and that key team members and all information will be present at the workshop.

Facilitation style 3: facilitation with full preparation

Preparation in this context involves several pre-workshop meetings with the client, key stakeholders and members of the design team. The facilitator may, as a part of the information gathering exercise, conduct a post occupancy evaluation of an existing facility occupied or owned by the client. During the pre-workshop stage the facilitator will begin to identify the issues and any mismatches in information received and to understand any hidden agenda. The membership of the team is fully discussed with the client, ideally to restrict multiple representations from one unit and add any stakeholders perceived to be useful to the workshop.

Having undertaken the pre-workshop activity described the facilitator will have greater confidence that the main issues have been identified and therefore the agenda for the workshop is appropriate, and that key team members and all information will be present at the workshop. The cost of the pre-workshop preparation is likely to be higher than in the preceding facilitation styles but there is more certainty of a worthwhile workshop.

4.14 Team composition

There are two fundamental points of view concerning the make-up of the value management team. The first, common in North America, is that the value management workshop team should be a totally independent review team overseeing the work of the design team, the team of record. The advantage of a totally independent team lies in its total independence. It has no preconceived ideas, brings no baggage to the workshop, can have a membership designed for the particular workshop and can be totally objective. The disadvantage lies in the fact that such a team can only be a reactive, backward-looking audit team. In other words the power of value management in proactively identifying a better way of proceeding throughout the project is largely lost. Additionally the independent team must convince the design team of the value of any suggestions to the extent that the design team will accept the liability of putting those suggestions into practice. Finally the independent team must be allowed time to familiarise itself with the project and to prepare a convincing case for the client and the design team.

The advantage of using the design team as the value management team – the situation most common in the UK – is the converse of the above. The disadvantage of using the design team is in the danger that it will not be objective or willing to change that which it has already decided upon. The counter to this is to use the value management process proactively and thereby the team plans the future of the project to the point where another workshop requires to be held.

4.15 Change management

Undertaking a project of any sort means that the project initiator is anticipating change. The reason for the project may be, for example, new legislation, a new commercial opportunity, a new policy for social enhancement, etc. The value management of the project may result in change not only to the client's core business – an invariable result of undertaking a project – but also to the project concept itself. Therefore change management is intrinsically linked to any value management output. The essential feature of change management is that it is future orientated and accepts that a current situation or operation is to be discontinued. The challenge is therefore to manage the move from situation A to situation B with the least physical and/or emotional friction.

If value management is carried out proactively, i.e. the functions of the project are discovered, recorded and finally specified by the project team through a VM process, then the issue of change is related only to the introduction of the project. If, on the other hand, value management is carried out reactively, i.e. the project specification has been completed and is being audited through a value management process, then change management relates to both the changes to the project specification and the introduction of the project to the client's core business. Obviously 'buy-in' by the team is easier if the team has been a party to the project specification through the VM process.

Acceptance of the change to an organisation by members of that organisation will depend largely on the organisational structure itself. In a military type of organisation, where management is through the exercise of authority and the imposition of sanctions, change will be directed. In another form of organisation comprised of largely autonomous units, such as a university, a hospital or an airport, change is likely at a unit level and buy-in becomes essential, particularly by the unit most affected by the change. In the latter case the team structure of a VM exercise allows representation from all stakeholder units.

The most difficult change management occurs when change results from inaccuracies in design caused by incomplete, unclear or ambiguous project information generated at any stage of the project process. This may result from, for example, the appropriate stakeholder's information not being incorporated at a particular stage in the development of the project. Easier to manage change occurs when a change brings benefits to all parties engaged on the project. Notwithstanding the type of change, management structures must be in place to deal with it.

Change management during the project comes in part from making clear, at the outset of the project, the consequences of change . This implies the development of a change control procedure which itself requires a frame of reference, a means of measuring and a method of remedial action. Change control therefore operates in three phases.

(1) *Planning the project timescale and activities, and identifying approval gateways.*
Approval gateways refer to points in the project when approval is given to

the scheme at that stage in its development. Effectively, after passing the approval gateway the project can only move forward. A change to the project that takes it back to a stage earlier than a previously passed approval gateway will necessitate passing through that gateway again. The identification of approval gateways is therefore fundamental to effective change management. Not only the major 'official' events such as business case approval, planning permission, etc. should be seen as approval gateways, but also the key events in the development of the project such as completion of the structural grid. After an approval gateway is passed a signed document recording the event should be inserted in the project execution plan (PEP). The project timescale, activities and approval gateways should be identified in sufficient detail to allow an effective frame of reference.

(2) *Recognising and identifying change.* Major changes can occur as a result of, for example, relatively minor changes to the client specification or to drawings. All changes and their consequences should be recorded and included in the PEP. Some change will be anticipated and the consequences planned for in the risk register. Identifying the change is undertaken by monitoring the frame of reference documentation. Simplistically, any deviation from the frame of reference not included in the risk register is a change.

(3) *Undertaking remedial action.* Before undertaking remedial action the reason for and source of the change should be identified, e.g. has a stakeholder changed their mind or made a mistake? Remedial action is in the following parts:
   (a) Measuring the time, performance and therefore cost consequences of the change and determining who will pay (if appropriate).
   (b) Reviewing the risk register and determining whether the change will increase the likelihood of an existing risk or give rise to a new risk.
   (c) Determining whether there is an alternative to the change.
   (d) Generating ideas for overcoming any impacts of the change on the elements in the client's value criteria.
   (e) Recording the change, obtaining signed authorisation and entering summary documentation in the PEP.

Change for the benefit of the project may be formalised into value management change proposals (VMCP) clauses within a contract. Changes suggested by organisations should aim to have the following characteristics:

- Reduce project programme.
- Not delay the project programme (subject to client's value criteria).
- Pay for abortive costs of redundant design and costs of the new design out of the savings generated by the VMCP (self-funding).
- Generate savings.
- Reduce risk.

Often incentive clauses are linked to such VMCP clauses.

From a team perspective the implementation of change management is obvious. Team members who benefit from the change will be supportive, those who are disadvantaged by the change will tend to be unsupportive and those who are unaffected by the change will tend to be ambivalent. The task of the value management facilitator is to reach consensus.

4.16 Conclusion

Value management and value engineering are facilitated team activities. This chapter has summarised current thinking in respect of groups and teams and outlined characteristics and methods for the facilitation of value management and value engineering teams. It was stated at the commencement that the choice of the individuals for the team is more related to job function that to individual trait, and this is an important factor for the facilitator to bear in mind.

More important, however, is the value management facilitator's strategy for the workshop. This chapter has presented the theory of teams and team dynamics, debating small teams (common in value engineering studies) and big team facilitation. That the latter are more challenging is without doubt, and the important thing here is to undertake the orientation and diagnostic work in sufficient detail that surprises, in terms of either team member behaviour or hidden agenda, do not come as a shock.

The facilitation models have been distilled into three forms. This is open to debate and is discussed again in Chapter 5 where the authors reflect on the theory and results of their own action research in order to evolve working models.

Chapter 5 draws together the structure, models and attitudes learned from observing international practice and forming them into a practice model. To do this the chapter draws on the material in this and the two preceding chapters.

4.17 References

1. Schein, E. H. (1980) *Organisational Psychology*, 3rd edn. Prentice-Hall, Englewood Cliffs.
2. Hellriegel, D., Slocum, J. W. & Woodman, R. W. (1998) *Organizational Behaviour*, 8th edn. South-Western College, Cincinnati, Ohio.
3. Cook, R. E., Hunsaker, P. L., & Coffey, R. E. (1994) *Management and Organizational Behavior*. Austen Press. Burr Ridge, Illinois.
4. Schein, E. H. (1980) *Organisational Psychology*, 3rd edn. Prentice-Hall, Englewood Cliffs.
5. Hellriegel, D., Slocum, J. W. & Woodman, R. W. (1998) *Organizational Behavior*, 8th edn. South-Western College, Cincinnati, Ohio.

6. Cook, R. E., Hunsaker, P. L., & Coffey, R. E. (1994) Management and Organizational Behavior. Austen Press. Burr Ridge, Illinois.
7. Hayes N. (2002) *Managing Teams: A Strategy For Success*, 2nd edn. Thomson Learning, Mitcham.
8. Male, S. P. (1991) *Competitive Advantage in Construction*. Butterworth-Heineman, Oxford.
9. Churns, A. B. & Bryant, D. T. (1984) Studying the client's role in construction management. *Construction Management and Economics*, 2 (1), 177–84.
10. Winch, G. (1988) The construction firm and the construction process: the allocation of resources to the construction project. In: *Managing Projects World-wide*, VII (eds P. Lansley & P. A. Halrow), pp. 967–75. E. & F.N. Spon, London.
11, 12. Hellriegel, D., Slocum, J. W. & Woodman, R. W. (1998) *Organizational Behaviour*, 8th edn. South-Western College, Cincinnati, Ohio.
13. Hunt, J. W. (1992) *Managing People at Work: a manager's guide to behaviour in organizations*, 3rd edn. McGraw-Hill, London.
14. Hunt, J. W. (1972) *The Restless Organisation*. Wiley, Chichester.
15. Hunt, J. W. (1992) *Managing People at Work: a manager's guide to behaviour in organizations*, 3rd edn. McGraw-Hill, London.
16. Hellriegel, D., Slocum, J. W. & Woodman, R. W. (1998) *Organizational Behavior*, 8th edn. South-Western College, Cincinnati, Ohio.
17. Belbin, Meredith R. (1981) *Management Teams: Why They Succeed or Fail*. Butterworth-Heinemann, Oxford.
18. Belbin, Meredith R. (1993) *Team Roles at Work*. Butterworth Heinemann, Oxford.
19,20 Hunt, J. W. (1992) *Managing People at Work: a manager's guide to behaviour in organizations*, 3rd edn. McGraw-Hill, London.
21. Adair, J. (1987) *Effective Team Building*. Pan, London.
22. Hellriegel, D., Slocum, J. W. & Woodman, R. W. (1998) *Organizational Behavior*, 8th edn. South-Western College, Cincinnati, Ohio.
23. Hunt, J. W. (1992) *Managing People at Work: a manager's guide to behaviour in organizations*, 3rd edn. McGraw-Hill, London.
24. Janis, I. L. (1972) *Victims Of Groupthink: A Psychological Study Of Foreign Policy Decisions And Fiascos*. Houghton Mifflin, Boston, Mass.
25. Janis, I. L. & Mann, L. (1977) *Decision-Making: A Psychological Analysis of Conflict, Choice and Commitment*. Free Press. New York.
26. Hunt, J. W. (1992) *Managing People at Work: a manager's guide to behaviour in organizations*, 3rd edn. McGraw-Hill, London.
27. Cameron, Esther (1998) *Facilitation Made Easy*. Kogan Page, London.
28. Male, S., Kelly J., Fernie, S., Gronqvist, M. & Bowles, G. (1998) *The Value Management Benchmark: Research Results of an International Benchmarking Study*. Published Report for the EPSRC IMI Contract. Thomas Telford, London.

5 Current Study Styles and the Value Process

5.1 Introduction

This chapter presents a synthesis of previous chapters as an overview of the value process including developments resulting from recent research. Different value study styles are introduced pursuing the argument that the role of the value manager is one of deciding on, structuring and delivering a study style tailored to a particular value problem, be it for a project, project programme, service or organisational function. Irrespective of the type of value problem it is postulated that the stages in its solution comprise three generic phases:

- Orientation and diagnostic phase
- Workshop phase
- Implementation phase

From a project perspective value management focuses on improving the business project and its relationship to delivering the technical project, whereas value engineering focuses on improving the delivery of the technical project to meet the aims and objectives set by the business project. This chapter discusses the relationship between value management and value systems, arguing that the former provides the mechanism to coalesce the latter.

The chapter combines findings from an international benchmarking study on value management[1] (referred to in this book as the benchmarking study) with observations from an action research programme[2] to shape a grounded theory[3,4] of value management. Generic observations are from over 200 studies conducted by the authors with clients, design teams, contractors and their supply chains under a range of different procurement systems. These studies also include working with organisations to introduce change into their structures, where the value management process has been particularly relevant in assisting them think through the issues.

5.2 The value process

The language of value management and value engineering normally involves discussion of the process, or the 'job plan'. While there are as many different job plans as there are authors on the subject, the benchmarking study identified a common underlying process reflected in the similarities, differences and limitations of the different job plans. The basic principles of these different job plans are shown in Fig. 5.1 with related activities at each stage in the process.

Following the publication of *The Value Management Benchmark* in 1998, the authors have increasingly focused on identifying value management study styles. These styles represent a combination of method, process and approaches to facilitation during the VM workshop phase. An important starting point for identifying study styles is the fact that the benchmarking study identified a series of intervention points in projects, with associated approaches to studies in terms of focus, value team composition, method and duration. The indicative durations given in the following text are the total time for a value management activity at that stage, i.e. if a full project briefing study had been completed then a following concept design study is unlikely to take more than four days with a half to one day workshop. At a concept design workshop it is only necessary to review the data that formed the brief and audit the drawn concept design against this data. Further, it should be stressed that while the workshop should be seen as a unitary event the other indicative days are not usually consecutive.

The authors have accommodated the above schematic of the process into a more simplified system for thinking about value studies (Fig. 5.2). The more simplified value process is set out below. It encapsulates the idea that value management is a change process and should be viewed in that light.

Figure 5.2 identifies three generic phases to a value study:

- *The orientation and diagnostic phase.* In this phase the value manager(s) and value team will be preparing themselves for the study, and the value manager(s) will meet with the commissioning client, project sponsor and key players who will be involved in the study, reviewing documents and possibly conducting interviews and briefings. The study style chosen by the value manager may also include understanding and structuring the value problem in detail. This may include exploring competing value problems, discussing possible solutions and exploring the way forward on completion of the workshop phase. The agenda for the workshop phase will be developed and the method and manner in which this will be conducted worked out. This phase will also set in train the process for, or as a minimum consider, the implementation of options and solutions developed from the workshop phase.
- *The workshop phase.* This is the stage where alternative and/or complementary views on the value problem will be brought together to explore and reach a way forward, hopefully through agreement. A workshop/study report will normally be produced, including an action plan to ensure that value solutions and options will be implemented in the post workshop phase.
- *The implementation phase.* This was targeted by the benchmarking study as one of the key areas where value management falls down. The authors have now adopted a variety of approaches to ensure this problem is minimised.

104 Method and Practice

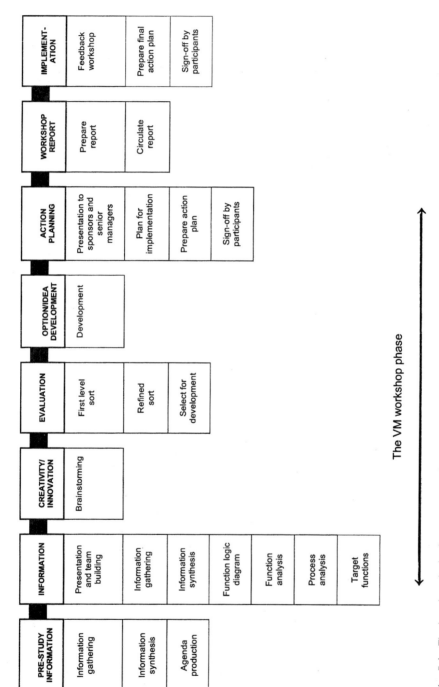

Fig. 5.1 The benchmarked value management process.
Source: Adapted from Male et al. (1998). *The Value Management Benchmark*[5].

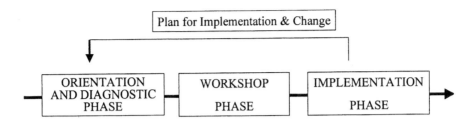

Fig. 5.2 The revised generic value study process.

Implementation meetings and workshops have been used. As a minimum, during the orientation and diagnostic phase an implementation strategy will be discussed with commissioning clients and wherever possible those responsible for implementation will be interviewed and identified in the action plan at the close of the workshop phase.

The three-phase generic value process will be adopted in this chapter as the frame of reference for discussing study styles.

5.3 Benchmarked study styles, processes and deliverables

Project intervention and value opportunity points

A study style is defined here as an outcome of the stage in the project life cycle at which a value study is carried out and the manner in which the process is conducted. This section reviews and develops further the styles identified in the benchmarking study. Figure 5.3 presents the benchmarked intervention points for value studies at stages of the project life cycle. Value opportunities are taken here to mean specific points in the project life cycle for value management interventions.

The studies conducted at value opportunity points are:

(1) Strategic briefing study.
(2) Project briefing study.
(3) Charette (C) – undertaken in the place of the studies at points 1, 2 and 4.
(4) Concept design workshop.
(5) Detail design workshop.
(6) Operations workshop.

The value opportunity points above are those most commonly found in practice. They are not seen as compulsory although some clients may insist that studies are conducted at these points due to their inclusion in standard operating manuals. The benchmarking research identified that they are the six probable points in the

106 Method and Practice

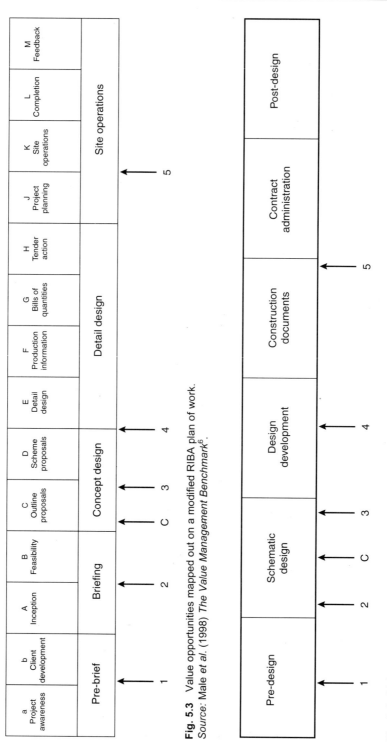

Fig. 5.3 Value opportunities mapped out on a modified RIBA plan of work.
Source: Male *et al.* (1998) *The Value Management Benchmark*[6].

Fig. 5.4 Value opportunities mapped out on an AIA design process.
Source: Male *et al.* (1998) *The Value Management Benchmark*[7].

project life cycle where value studies are beneficial. The research also identified that value opportunities can be taken at single opportunity points in the project life cycle or at multiple intervention points. Each opportunity point will be discussed in turn.

Strategic briefing study: benchmarked value opportunity point 1

The strategic briefing study is concerned with identifying the broad scope and purpose of the project and its important parameters. The focus is on articulating strategic needs and wants, the role and purpose of the 'business project' for the client organisation clearly expressing the reason for an investment. Hence the intent of the study style is to answer the questions why invest, why invest now and for what purpose? A strategic briefing study describes clearly and objectively the 'mission of the business project' and its strategic fit with the corporate aims of the client organisation. These corporate aims are explicit in terms of commercial objectives and usually implicit in terms of cultural values. The commercial objectives and cultural values combined form the value system of the client organisation, the client's value system. This value system along with the client's methodology for total quality management should be overtly expressed as a part of the strategic brief. The structure and operational methods for determining the client's value system is described later in this book but is the value criterion against which all business project decisions are judged.

An important deliverable is the output specification explaining clearly what is expected of the 'business project'. This will include establishing the outline budget and programme. The strategic briefing study explores a range of options for delivering the 'business project', one of which could be the creation, refurbishment or renewal of a physical asset or assets as a corporate resource. 'Optioning' could involve developing and investigating non-physical asset alternatives. The strategic briefing study will structure information in a clear and unambiguous way to permit the 'decision to build' to be taken in the full knowledge of all the relevant facts. On completion of the strategic briefing study, the decision to build can proceed with confidence, given that all relevant issues and options have been addressed and explored, and alternatives examined.

One powerful technique useful when nearing completion of this study style was highlighted during the authors' research into architectural programming in North America[8]. An eminent architectural programmer in Canada used the analogy of pressing a 'go button'. The question he would ask of senior managers in client organisations would be:

> 'Are you sure you are ready to push the Button and unleash the full resources of the construction industry, the destinies of hundreds if not thousands of people and affect the fortunes of numerous firms and organisations, including your own?'

Interestingly, asking a simple question such as this returned many of his clients to the analysis stage of the strategic briefing study.

Indicative techniques during the orientation and diagnostic phase of the study would include:

- Interviews.
- Stakeholder mapping.
- Document analysis.
- Questionnaires.
- Post occupancy evaluation of a similar facility or of the facility under discussion in the case of refurbishment and adaptation projects.
- Site Tour.

Indicative techniques during the workshop phase of the study would include:

- Presentation and team building.
- Issues analysis, group, theme, prioritise.
- Client value system.
- Stakeholder analysis.
- Strategic time line.
- Project driver analysis.
- Time/cost/quality analysis.
- Functional analysis.
- Function logic diagram.
- Brainstorming alternatives.
- Evaluation and development.
- Presentations from working groups.
- Plan for implementation.
- Prepare action plan.
- Prepare and circulate study report from which the strategic brief would be developed.
- Sign off workshop report by participants.

Deliverables in the strategic brief document would include:

- The mission statement for the business project – a clear statement of why invest now.
- The project context.
- The client's value system and particularly how the success of the project will be measured.
- Organisational structures for project delivery.
- Overall scope and purpose of the project.
- High level risks.
- Programme, including phasing.
- A global capital expenditure budget and any cashflow constraints.
- Initial options for inclusion into a procurement strategy.
- Targets and constraints on operating expenditure and other whole life costs.

- An implementation plan. This would include the decision to build or factors to be considered in the decision to build.

The benchmark study indicated that the study would typically take between four and seven days to complete, with the workshop phase taking half to one day of that time scale, although this can vary depending on the size and complexity of the project and its sensitivities. Participants are likely all to be at senior levels within the client organisation and it is not uncommon for teams to be in the size range of 10 to 20 people or more, involving 'big team' facilitation skills during the workshop phase.

To summarise, the primary purpose of this study is to develop a strategic brief which describes in business language the reason for an investment in a physical asset, its purpose for the organisation and its important parameters. Anybody reading the strategic brief should be able to understand why an organisation has decided to invest in a physical asset or assets and pursue no other strategic options that might compete for the same investment resource at that time. In the context of the UK government OGC's Gateway review process reviewed in Chapter 2 this study would be at Gateway 0: strategic assessment. The workshop report is an audit document describing the reasons for decisions made at this point.

Project briefing study: benchmarked value opportunity point 2

The project briefing study focuses on delivering the 'technical project'; that is, the construction industry's response to client requirements expressed in the strategic brief. The project brief translates the strategic brief into construction terms, specifying performance requirements for each of the elements of the project. If it is a building project this will also include spatial relationships. An outline budget will also be confirmed if a strategic briefing study has been undertaken, and developed if not.

Indicative techniques during the orientation and diagnostic phase of the study would include:

- Interviews.
- Stakeholder mapping.
- Document analysis.
- Benchmarking information from similar projects.
- Questionnaires.
- Post occupancy evaluation of a similar facility or of the facility under discussion in the case of refurbishment and adaptation projects.
- Site tours.

Indicative techniques during the workshop phase of the study would include:

- Presentation and team building.
- Issues analysis, group, theme, prioritise.

- Client value system.
- Stakeholder analysis.
- Strategic time line.
- Project driver analysis.
- Time/cost/quality analysis.
- Function logic diagram.
- REDReSS.
- Process flowcharting.
- Functional space analysis.
- Spatial adjacency analysis.
- SWOT.
- Brainstorming alternatives.
- Evaluation and development.
- Presentations from working groups.
- Plan for implementation.
- Prepare action plan.
- Prepare and circulate study report from which the project brief would be developed.
- Sign off workshop report by participants.

The project briefing document would include:

- A summary of the relevant parts of the strategic briefing document.
- The aim of the design. This would include priorities for project objectives.
- The functions and activities of the client, including the structure of the client organisation and the project structure for delivering the project.
- The site, including details of accessibility and planning.
- The size and configuration of the facilities.
- The skeletal project execution plan or update in the case of it being independently prepared by the project manager.
- Key targets for quality, time and cost, including milestones for decisions.
- A method for assessing and managing risks and validating design proposals.
- The procurement process.
- Environmental policy, including energy.
- Outline specifications of general and specific areas, elements and components in output terms.
- A cost centred budget for all aspects of the project including all elements of the construction project.
- Options for environmental delivery and control.
- Servicing options and specification implications, e.g. security, deliveries, access, work place, etc.
- Key performance indicators for each stage of the project.

The benchmark study indicated that the study would typically take between four and eight days to complete, with the workshop phase taking one to two days of that time scale, although this can vary depending on the size and complexity of the project and its sensitivities. Participants are likely to be senior representatives from within the client organisations, and design and project management team representatives. Again, it is not uncommon for teams to be in the size range of 10 to 20 people or more, involving 'big team' facilitation skills during the workshop phase.

To summarise, the primary purpose of this study is to develop the project brief, which describes in technical terms the technical project to be delivered by the construction industry that will respond to and/or deliver the strategic brief. The former provides the basis on which design can proceed. In the context of the UK government OGC's Gateway review process reviewed in Chapter 2, this study would follow Gateway 1: business justification. The workshop report is an audit document describing the reasons for decisions made at this point.

Concept design study: benchmarked value opportunity point 3

The concept design study is a value review of the initial plans, elevations, sections, outline specifications and cost plan of the proposed built asset. The study will focus on validating the concept design or assisting the further development of design options and improvements. The assumption is that the client has agreed the project brief, although this would be tested as part of the study process. A good starting point for considering a concept design study is that, for most projects, the design has reached the point of seeking detailed planning permission.

Indicative techniques during the orientation and diagnostic phase of the study would include:

- Interviews.
- Document analysis.
- Benchmarking information from similar projects.
- Site tour.
- Questionnaires.
- Post occupancy evaluation of a similar facility or of the facility under discussion in the case of refurbishment and adaptation projects.
- Facilities walk-through.

Indicative techniques during the workshop phase of the study would include:

- Presentation and team building.
- Issues analysis, group, theme, prioritise.
- Client value system.
- Stakeholder analysis.
- Strategic time line.
- Project driver analysis.
- Time/cost/quality analysis.

- Function logic diagram.
- Process flowcharting.
- Spatial adjacency analysis.
- Functional space analysis.
- Major element function analysis and diagramming.
- REDReSS.
- SWOT.
- Brainstorming alternatives.
- Evaluation and development.
- Presentations from working groups.
- Plan for implementation.
- Prepare action plan.
- Prepare and circulate study report.
- Sign off workshop report by participants.

The outputs for the concept design study would include:

- A statement of the direction of the design.
- The project execution plan or update in the case of it being independently prepared by the project manager.
- The procurement strategy and the options explored for this.
- Key milestones.
- Key performance indicators.
- Important risks, including a risk management strategy.
- A detailed cost plan and a detailed budget.
- A schedule of activities.
- The site layout and access, including the identification of ground conditions and any planning constraints.
- Dimensioned outline drawings and an outline specification for all systems.

The benchmark study indicated that the study would take typically between four and eight days to complete, with the workshop phase taking one-and-a-half to three days of that time scale, although this can vary depending on the size and complexity of the project and its sensitivities. Participants are likely to be senior representatives from within the client organisations, and design and project management team representatives. Again, it is not uncommon for teams to be in the size range of 10 to 15 people during the workshop phase. In the context of the UK government OGC's Gateway review process reviewed in Chapter 2, this study would be at Gateway 2: procurement strategy. This is a useful point at which to take an objective view of the proposed procurement process, incorporating the reasons for the decision and the actions in the workshop report and action plan.

To summarise, on completion of the concept design study the design team may develop further options identified during the study or continue with normal design development in the full knowledge that the team has explored fully the design development to date and confirmed its acceptability to the client.

The charette: benchmarked value opportunity point C

The charette is a hybrid study. It is an audit of the project brief and is often undertaken once the concept design is complete. It audits the concept design against the strategic brief and project brief. The benchmarking study highlighted that in North America this study is often referred to as being undertaken at 10% design. The charette is commonly the first study undertaken on a project. It implies that the client has reached the decision to build, completed the project brief, appointed a design team and then undertakes a value management study. The study is wide ranging, incorporating the previous three studies discussed above. The study focuses on validating the project brief, and frequently the concept design, to ensure that both conform with and fulfil the client's value system. A primary purpose of the charette is to ensure that the client value system is overtly described and understood.

Indicative techniques during the orientation and diagnostic phase of the study would include:

- Interviews.
- Stakeholder mapping.
- Document analysis.
- Benchmarking information from similar projects.
- Questionnaires.
- Post occupancy evaluation of a similar facility or of the facility under discussion in the case of refurbishment and adaptation projects.
- Site tour.

Indicative techniques during the workshop phase of the study would include:

- Presentation and team building.
- Issues analysis, group, theme, prioritise.
- Client value system.
- Stakeholder analysis.
- Strategic time line.
- Project driver analysis.
- Time/cost/quality analysis.
- Function logic diagram.
- Functional analysis of space, elements and components.
- VE element function diagramming.
- Process flowcharting.
- Spatial adjacency analysis.
- REDReSS.
- SWOT on the design.
- Brainstorming alternatives.
- Evaluation and development.
- Presentations from working groups.

- Plan for implementation.
- Prepare action plan.
- Prepare and circulate study report which validates or amends the project brief.
- Sign off workshop report by participants.

The outputs from a charette would be a combination of the deliverables identified above in studies 1 to 3. The benchmark study indicated the study would typically take between four and eight days to complete, with the workshop phase taking two to three days (16–36 hours) of that time scale. Participants are likely to be senior representatives from within the client organisations, and design and project management team representatives. Again, it is not uncommon for teams to be in the size range of 10 to 15 people during the workshop phase. The authors' experience is that the charette is a common type of value management exercise undertaken on construction projects where value management is a single event.

To summarise, on completion of the charette study the client value system would have been made explicit, the project brief would have been validated and any outline designs audited against the client value system, strategic brief and project brief. The design team would develop further options identified during the study or continue with normal design development in the full knowledge that the team has fully explored the strategic and project briefs and confirmed their acceptability to the client.

Final sketch design/scheme design workshop: benchmarked value opportunity point 4

Once the client project manager has 'signed off' the concept design, the project team should begin the development of the final sketch design and specification of the performance requirements for elements of the facility. The final sketch design should freeze as much of the design as possible, defining and detailing every component of the construction work. It should identify further risks associated with the project and outline proposed action if they arise, assess the quality requirements and define how success will be measured. The focus of the study moves from the strategic and client organisation to the technical solution of the concept design and involves value engineering the element function and whole life performance relationships.

Indicative techniques during the orientation and diagnostic phase of the study would include:

- Document analysis.
- Benchmarking information from similar projects.
- Site tour.

Indicative techniques during the workshop phase of the study would include:

- Presentation and team building.
- Review of the client's value system.
- Review of the brief.
- Review of the concept design.
- Review of the time line and project programme.
- Design drivers.
- Time/cost/quality analysis.
- Functional analysis of space, elements and components.
- VE element function diagramming.
- SWOT on the design.
- Function component analysis.
- Pareto analysis – histogram of cost.
- Cost vs. value.
- Identify mismatches.
- Brainstorming alternatives.
- Evaluation and development.
- Presentations from working groups.
- Plan for implementation.
- Prepare action plan.
- Prepare and circulate study report.
- Sign off workshop report by participants.

The outputs from the detail design study should include:

- A statement of scheme design.
- Update of the project execution plan.
- Key milestones and targets.
- Performance measures.
- Location of site, information on planning approvals and other detailed permissions agreed.
- Dimensions of spaces and elements provided.
- Performance specifications for environmental systems and services.
- Further risks and a risk management strategy.
- The cost plan.
- Proposals for the maintenance and management of the completed facility.

The benchmark study indicated that the study would typically take between four and nine days to complete, with the workshop phase taking two to five days of that time scale, again depending on the complexity and size of the project. Participants are likely to be senior representatives from within the client organisations, and design and project management team representatives. Teams will be in the size range of 10 to 15 people during the workshop phase. As discussed in Chapter 3, large technical team workshops to consider all project elements at a single sitting are being superseded by cluster group workshops considering clusters of elements. While this loses the benefit of all specialists seeing the impact of their work on all elements it has the advantage of involving small groups of specialists considering those elements of direct relevance. A small cluster workshop considering substructure, frame and upper floors may meet for a half to one day to consider this cluster in isolation. The common denominator at the workshop would be the design team leader, project manager and construction manager.

To summarise, on completion of the final sketch design study the client value system would have been made explicit within design development. The design team would continue to develop element component designs for whole life performance.

Operations workshop: benchmarked value opportunity point 5

The Operations study converts design into component and constructional operation sequences. It is undertaken at the point at which construction work is about to commence.

Indicative techniques during the early stages of the study would include:

- Interviews.
- Exploration of contractor's viewpoint and strategy.
- Assembly of production drawings and contract programme.
- Preparation of histograms from tender information.
- Identification of major work packages.
- Making explicit change management procedures.

Indicative techniques during the workshop stage of the study would include:

- Presentation and team building.
- Issues analysis, group, theme, prioritise.
- Risk analysis.
- Review contract programme.
- Supply chain analysis.
- Functional specification.
- Component analysis.
- Pareto analysis – histogram of cost.
- Pinch point and committal point analysis.

Outputs from the Operations Study would include:

- A statement of the extent of design consistent with the procurement route.
- The project execution plan.
- Key milestones and targets.
- Key performance indicators.
- A supply chain diagram.
- Pinch points or gates in project development, which have a strategic or tactical impact on following work packages.

- Identification of key work items to be targeted for specific technical workshops.
- Risk management plan.

The exact definition of this stage will depend upon the procurement method adopted. The operations study will introduce supply chain and technical development issues. It should update the risks associated with the project and appraise the proposed action identified earlier.

The benchmark study indicated that the study would typically take between two and six days to complete, with the workshop phase taking one or more days if a series of workshops is programmed, again depending on the complexity and size of the project. Participants are likely to include contractor's production planning, purchasing, project management, supplier and/or subcontractor representatives. Depending on the procurement route, client, design team and client consultant project managers may be present. Teams will be in the size range of six to ten people during the workshop phase.

To summarise, on completion of the operations study site operations would commence, with a series of technical supply chain workshops also programmed as part of the action plan.

5.4 Other study styles

The value management process is a flexible group decision support system that has been used by the authors in a variety of different situations. This section discusses other study styles that build on the generic process used for construction projects. Examples include:

- *Single project versus project programme studies.* The preceding section has generally outlined approaches relevant for different stages of individual projects. However, the same process can be used to develop the strategy for a programme of projects, where the focus will be on identifying the objectives of the overall programme, individual projects within the programme and how they fit together holistically. Programme level VM studies will also focus on understanding and resolving competing objectives and resources between projects making up the programme.
- *Organisational change studies.* The value process and functional analysis has proved a very powerful technique in assisting the authors to help organisations restructure divisions, departments and teams. Organisational change studies have involved assessing where the organisational unit under study is now and where it wants to be in the future, perhaps over different time frames. A prioritised issues analysis assists in targeting the important change issues, what is critical for success and where the important risks might lie. Functional analysis can be used to describe the essential functions that the

new organisational unit must perform in the future, with skill and resource requirements being worked out to meet the functional structure developed by a value study team. Normally, a migration strategy will also need to be developed to articulate how the organisational unit will get from its current state to the future state described by the function diagram. Underlying the process described is a continued questioning of which functions will add value to the organisational unit in the future. The process has also been used to implement a risk framework within a firm across all of its subsidiaries, again using functional analysis to define the reason why the risk framework needs to be implemented and its purpose in the organisation.

- *Facilities programming studies.* The authors have conducted a range of studies that have developed the technical or project brief for a project and/or included defining the appropriate procurement route. The focus of this type of study is to ensure that business and technical projects are in alignment and that the technical/project brief reflects this alignment in the specifying performance requirements for elements of a building. A full shadow team comprising architect, M&E, structural, QS and contracting members are used to brief the project with the client and end users, including defining spatial adjacencies, elemental performance specification, procurement route and budget. The term facilities programming is used in this instance to describe this type of study since the technical/project brief leaves the client totally free to chose the procurement method and the best supply team to deliver the project. Post occupancy evaluation studies can also provide a good source of information into the briefing process since cross project learning is available. The facilities programming study goes beyond a normal project briefing study, which is usually conducted with an existing team of record. The former combines a strategic and project briefing study, often with procurement strategy options included as part of the analysis, using an independent team working alongside the client.

- *Project audits and value-for-money studies.* The value process can be used to undertake project audits and/or value-for-money studies. Where the authors have used this, a full shadow team has been utilised to provide recommendations and opinions covering a broad spectrum of expertise on whether the service/project to that stage is providing value for money. This approach is similar to North American-style VE studies[6]. The approach also has considerable merit for undertaking best value reviews under the UK government's Local Authority initiative.

- *Procurement studies.* These have been wide ranging in form but there have been two distinct characteristics to them: first, those that involve the client and its advisors in working out the best procurement option for a project; or second, bid conferences, where the value process has been used to assist contractors to develop a bid strategy under a particular procurement route and subsequently considering supply chain procurement issues. Each will be discussed in turn.

☐ *Client led procurement studies.* With these studies the value process is used to develop and understand in detail the client value system. Subsequently this will be translated into an analysis that attunes the client value system to the appropriate procurement route, including decisions on attitudes to risk and its allocation. While the authors have conducted procurement studies in their own right, it is not uncommon for procurement to be addressed as part of another type of value study as indicated above.

☐ *Bid conferences.* These have taken a variety of forms:

Traditional design and build. Here the value process is used comprehensively to understand the bid documentation, design and construction interactions, risk and the client value system, usually expressed through building a function diagram. Tender development teams tackle areas that need to be developed further for the bid. The advantage of using the process in this way is that bid documentation can be broken down in a structured, comprehensive way, with value engineering options also being developed as as part of the process. One disadvantage of using the value process in this way is not having direct access to the client in a competitive bidding situation to fully understand the client value system. However, there is usually sufficient information within the bid team to be able to construct a robust and comprehensive understanding of the client value system, albeit a second guess.

PFI, prime contracting and NHS Estates Procure 21 procurement systems. The authors have used the value process to assist teams at either pre-qualification, invitation to tender, best and final offer, preferred bidder or post contract stages. For one particular project at PFI preferred bidder stage, a combined value management and value engineering exercise was undertaken at risk to the consortium. It involved the end users assisting the consortium to finalise the scheme and drawings prior to contract close. The value process has also been used at post contract award stage to ensure that the project organisation has been set up appropriately for the remainder of the contract. The key benefits of using the value process under these forms of procurement are that it speeds up project learning, defines the important issues early on in the bid process and can also save a bid team time by increasing cohesion, enhancing commonality of thinking across the supply chain team, assisting forward planning of the bid process and the project, understanding client drivers and identifying critical success factors. Using the value process also enhances a partnering ethos and assists embedding VM and risk management early into supply chain clusters, where the procurement system requires this approach. Typically, clustering strategies will also be developed using the value process.

■ *Partnering studies.* The authors consistently approach partnering from a value process perspective. Partnering comes in many guises and is understood to be anything from good teamwork within an otherwise traditional contractual framework, to a situation where there is no formal contract and full partnering exists. Within this spectrum there are recognised to be two primary partnering types; project partnering and strategic partnering or alliancing. Project partnering is a partnering framework for a single project. The partnering may evolve at an early stage and incorporate a selection procedure or may be instituted after a traditional tendering process and after a contract has been signed. In strategic partnering or alliancing the partnering agreement exists across organisations and is independent of specific projects. In either case it is common in construction for partnering contracts such as PPC 2000 to be used or partnering procurement approaches that require it as an overlay on standard forms of contract.

The value approach involves understanding and respecting each party's issues and objectives whether commercial or social. It is often difficult for public sector clients to understand, accept or respect the fact that a commercial organisation has a culture based on profit and is entitled to earn a profit, and for private sector organisations to understand that the public sector client has a duty to procure socially orientated services representing best value. An understanding of the cultural issues of each party is fundamental to effective partnering which is generally characterised by trust and openness between the parties, a robust system for sharing gain and pain, and a recognition that tasks should be undertaken by the best person for the job irrespective of the organisation to which that person belongs.

Normally, the full gamut of VM tools and techniques will be used in a series of partnering workshops to set up the ethos from the outset, and define the charter and the important components of the partnering memorandum of agreement. This is successfully achieved using issues analysis to understand all parties' issues as indicated above. Function diagramming is used to derive the partnering mission and the level 1 functions the components of the partnering memorandum. Typically, partnering workshops will combine process analysis, where appropriate, and define incentivisation, risk management and risk sharing systems.

- *Strategic studies.* The authors have also adapted the value process for use on complex studies that combine elements of both corporate and project strategy. These types of studies have included rationalising plant locations, improvements to industrial processes where temporary construction activity may impinge directly on the delivery of industrial products, and the development of business cases to demonstrate the benefits of new ways of working compared to existing methods.

Figure 5.5 draws together the different study styles that the authors have adopted in practice.

To summarise and conclude, value management is a service with three primary core elements: a value system, a team-based process, with functional analysis promoting understanding. It has three generic processes to it:

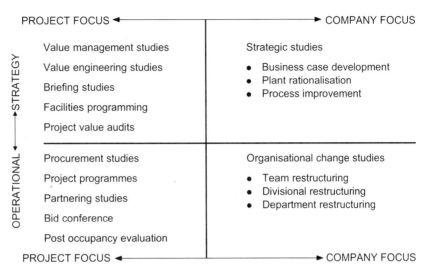

Fig. 5.5 Different study styles.

- The orientation and diagnostic phase.
- The workshop phase.
- The implementation phase.

The role of the value manager is to structure a study strategically and tactically to take account of these phases within it. The skills of the value manager comprise the ability to understand a value problem, structure a process to bring value systems together and introduce improvements subsequently. Value management is, therefore, a change orientated process and needs to be treated, designed and delivered as such.

5.5 Observations from practice

The following sub-sections present observations from the action research programme discussed earlier.

Who initiates a value study and who commissions it?

The authors have found that the approach of the commissioning agency is an important issue for the success of a study. A study may be a requirement within a set of standard project operating procedures, in which case it forms part of an ongoing system of value studies. Other instances may include either a senior manager or a project manager within the client organisation commissioning a study because a project has run into difficulties, the most common reason being that the project is over budget. Other examples for initiating a study include a client wishing to use the approach proactively as part of the project development process, or VM/VE forming a mandatory part of a procurement route, such as within the UK government's OGC Gateway process reviewed in Chapter 2. Sometimes the person that initiates the study may be different from the person commissioning it.

Behind the foregoing is the requirement to establish the scope of the study, the constraints, both absolute and movable, the objectives of the study and the deliverables. It is also important to establish the degree of independence that the value manager has within the study. This will determine the degree of latitude and authority that a value manager has throughout the process and the extent to which competing value systems during the workshop phase can be 'challenged' in moving towards integration.

The job plan

Value management and value engineering have traditionally been built as a process around the job plan, considered to be a good and effective decision making process. The benchmarking manual[1] brought together value study processes, adding to them to reflect discussions forming part of the research. The benchmarked value process included implementation, previously missing from all job plans. This was raised unanimously as an important issue by all research collaborators.

The job plan, especially during the workshop phase, implies a sequential process and strong proponents of it argue that it should be adhered to strictly. However, practice has demonstrated clearly that operating sequentially can sometimes hinder flexibility and innovation, and an unquestioning application may not be in tune with team dynamics during the workshop phase. For example, in certain studies the authors have discerned that a value team may be ready to move to the development stage ahead of the normal sequence; the team wanting to move ahead, being aware of what they need to do. The decision to move out of sequence is a matter of judgement on the part of the facilitator but has to take account of what has been achieved and what is to be achieved during the workshop phase. It also has to take account of the team's readiness to proceed out of sequence, team relationships and the advantages and disadvantages of doing so. In instances where the sequence has not been adhered to it has proved beneficial and the value team has subsequently returned to earlier stages of the job plan to complete activities normally considered earlier.

To conclude, it is best to consider the value process identified in Fig. 5.4 as an ideal sequence that promotes effective decision making in value teams. However, effective, innovative, decision making may require adjustments to the process and the situation may dictate when this should happen. This is a matter of workshop tactics.

Information gathering: the orientation and diagnostic phase (O&D)

Blyth & Worthington[10] have described the value management process as an 'explosive intervention' into a project. The authors would agree. With this in mind, the orientation and diagnostic phase (O&D) of a VM study is crucial. It is the only opportunity to fully understand the project, the service, the participants, the stakeholders and the 'value problem' – stated, perceived or real. The orientation and diagnostic phase is about information gathering, synthesis, seeking clarification and deciding the appropriate way forward for a value study. For example, value studies conducted by the authors have sometimes revealed that the stated 'value problem' is not the one that emerges finally for solving as a study progresses. With this in mind, as indicated above, the authors have now adopted and included the orientation nomenclature used by Miles[11] to also explain part of the emphasis during this phase of a study. The 'orientation' term emphasises the orientation by the value managers to the value problem. The 'diagnostic' term emphasises that value managers could find themselves uncovering competing value problems, value systems and differing ways to approach the workshop phase of a study, or even the orientation component itself.

The benchmarking study also highlighted that different value managers will have different approaches to this phase of a study. These approaches are discussed in more depth in Chapter 4. The issue is one of how much preparation should be undertaken. Some value managers argue none; others argue some, mainly focused around getting out and prioritising issues. The authors prefer to adopt full preparation during the O&D phase wherever possible as it is important to know about the 'value problem' being tackled in order to design for and provide for a good outcome from the study. The workshop phase can be planned better and work more effectively by discovering competing contenders for the 'value problem'. These will have to be explored subsequently during the workshop process.

One of the debates that emerged during the benchmarking study was the extent to which attendees for the workshop phase should be provided with briefing information prior to the workshop. The diversity of views is best encapsulated via the following:

- A major summary document should be produced detailing information gathered.
- A ten-page briefing document should be sent out in advance capturing the important elements of the study, summarising the information gathered, the agenda and how the value process will operate.
- An agenda should be the only information forwarded to attendees.

The authors have found that, in practice, if too much information is sent out it will not be read. In general, the value process designed and adopted by the authors does not necessitate sending out much information in advance. A detailed agenda for the workshop phase will often suffice, especially given that the authors normally anticipate interviewing key workshop attendees in advance. However, much will depend on the nature of the study in question. In instances where attendees at the workshop may have substantive differing views on the value problem, a detailed interim report has been produced prior to the workshop phase encapsulating generic competing perspectives. This prepares attendees for the fact that the workshop may be difficult, that they may find their positions challenged and that they may have to change their stance. However, this is a matter of study strategy for the workshop phase and depends on the value problem to be addressed. Use of this approach has to be thought through carefully as the report may entrench positions further, or it may soften them; a matter that needs careful discussion with the commissioning body for the study.

There have been instances where initial thoughts were that this is an appropriate approach and as the O&D phase has developed it has been decided not to proceed with the use of an interim report. However, in other instances as the phase has developed this has come to be seen as the only approach to move entrenched positions. The focus of this strategy is concerned with expectation structuring by the value managers during the workshop phase.

The O&D phase, as well as being associated with information gathering, will be used to agree participants for the workshop phase of the study. Study participants may be pre-selected by the commissioning client; the value manager may have some input into the selection process or may have a free hand and agree attendees in consultation with the client. The ACID test is a useful tool for determining the identity of attendees. It could be the case, where interviews and/or documentation analysis are undertaken, that additional participants come to the fore. The authors' preference is to have a degree of influence over who should attend, not least to restrain numbers and keep costs down.

The role of interviews during the orientation and diagnostic phase

Wherever possible the authors prefer to interview key attendees for the workshop phase before this takes place. Interviews may not always be required or possible; however, they are particularly useful for complex value problems or those that are highly political or sensitive. In such situations, the authors would stress the importance of interviews to the commissioning body. Interviews can also be useful when there is a high degree of apprehension about the value study itself, and about the workshop in particular. Interviews can also be useful for building commitment to the value study. The use of interviews in the foregoing situations means that interviewees must be assured that all discussions will remain confidential between them and the value managers. Thus, interviews perform a number of functions during the O&D phase:

- ■ First, competing and legitimate views from a value team on a way forward can be highlighted, hidden agendas uncovered, and also possible solutions explored. The authors have often found parts of the jigsaw for solving a particular 'value problem' exist in the minds of key people already. These partial solutions can often be uncovered during interviews and require the workshop phase to make them explicit, crystallise them and then design an appropriate way forward as part of the workshop process. Interviews in this situation provide an essential overview of the pieces of the jigsaw and how they may fit together.

- Second, depending on the purpose of a value study, the authors have also used interviews to raise with key attendees that during the workshop phase they may need to redefine their position or seed that change is required from them.
- Third, the authors will also use interviews to talk through the workshop process, answer any questions and brief attendees on what might be required of them. This has been exceptionally useful on politically sensitive projects.
- Finally, the interview process can also highlight additional expertise that may exist outside of the existing team or organisation and that may be required for the workshop phase.

The use of site tours during the orientation and diagnostic phase

The authors have found site tours a very useful technique during the O&D phase. A site tour makes the location for a project 'real'. The value managers can understand the layout of a project, any site constraints, local community issues, etc. It is also a good adjunct to interviews undertaken before the workshop phase. Often remarks can be made during a site tour that may not normally come out during formal interviews, or are reflected in documentation but are essential for gaining a full understanding of a value problem.

The client and other value systems

An important component of the O&D phase is to build up a picture of the client value system. This will be derived from documentation analysis, interviews with key stakeholders, and views expressed during any site tour. The O&D phase will also be used to understand other value systems present in the value study, for example designers, constructors, cost consultants and other stakeholders. This will provide the value managers with an insight into the extent to which value systems are aligned or diverge, and where common ground and differences may lie. It assists in preparing the agenda for the workshop phase of the study and in deciding on tools and techniques to be used during the workshop.

The workshop agenda

The agenda for the workshop phase will be developed during the O&D phase. It is an outcome of that process and is a sequence of operations normally expressed as tools and techniques. At times, the authors have made agendas explicit and detailed to forewarn teams of what is ahead of them and to ensure that a team comes prepared, knowing what the scope and structure of the workshop will be. As discussed above, it is often necessary to use an alternative tool or technique to that described in the agenda where the flow and direction of the workshop is seen or felt to be changing. Depending on the sensitivity of studies and value problems, the authors have sometimes left agendas open such that issues are allowed to emerge during the workshop process and prevent unwarranted expectations ahead of the workshop phase, potentially to the detriment of the study. Again, this forms part of a study strategy and expectation structuring.

Workshop phase

The workshop phase is an important part of the value management process. However, it can be expensive and time consuming, and can lack purpose and direction if not properly planned and executed. It is also the period when value systems may:

(1) Align.
(2) Re-align.
(3) 'Collide'.
(4) Be reconciled.
(5) Be integrated.
(6) Diverge.
(7) Coalesce into a common understanding.

The authors have experienced all of the above, often many within the space of one workshop depending on the nature of the project and value problem under study. For this reason the workshop phase has been described as a 'pressure cooker', with due respect given to the power of value systems, culture, paradigms and perspectives present within the workshop process. The authors now tend to view the workshop process as falling into four major groups of activities as demonstrated in Fig. 5.6. This provides a useful guide for ordering and

Presentations Information sharing Prioritise information Commence team building	Evaluate solutions Develop winning solutions
Back-to-basics Identify value mismatches Brainstorm solutions	Present and agree winning solutions Develop action plan

Fig. 5.6 The major workshop components.

structuring the workshop process, deciding on tools and techniques and planning team procedures. Equally, it is a useful guide for deciding if the value process should proceed sequentially or if the value problem necessitates re-ordering the sequence of activities.

The workshop phase involves facilitating a team, understanding team dynamics and controlling the process to ensure a successful outcome. However, by its nature, a workshop may involve the value manager retaining, exerting or giving up control of the team. This is a matter of tactics within the overall strategy for the workshop and the study in general. The value manager becomes a process manager during the workshop phase, using the skills of facilitation to move the process forward.

The authors have experimented with different forms of agendas, and over time have learnt that they should not be too detailed but should provide sufficient information to structure the process. If workshop agendas are too detailed, perhaps with timings identified very tightly, attendees can become more concerned about the agenda and progress than about the content of the workshop. The important point is that agendas provide a structure to the workshop, permit a broad framework to be highlighted to the team and, if the workshop process dictates, can be adjusted to suit evolving circumstances. The authors now anticipate agendas to be modified. Workshop processes inevitably bring things into the open that necessitate adjustment to process and timing. In this sense, the skills of the value manager are about strategy and tactics. The workshop is designed and used strategically in the context of the value study. Tactics are used during the workshop process, usually involving modifications to the agenda, changing tools and techniques and adapting team processes to adjust the workshop dynamics as the situation dictates. Figure 5.7 demonstrates the linkages between major workshop facets and tools and techniques.

Observation has again highlighted that there are underlying psychosocial processes at play during various stages of the workshop. The categorisations of opinions, beliefs, attitudes and values are useful in this context to understand workshop dynamics. Opinions operate at the individual's psychological boundary whereas beliefs, attitudes and values operate much deeper in the psychological core of the individual. The manner in which the workshop is structured and the tools and techniques used will tap into these elements of personality at different times and in different ways.

The authors have designed their approach to workshops to take account of these psychosocial dynamics. For example, an issues analysis is a workshop opening technique. It allows individuals to get out into the open anything that they want to raise about the value problem under investigation. The issues analysis is both visual and wide ranging. It permits a team to talk openly and allows them to relax with each other. However, following on from an issue analysis brainstorming session the value managers will subsequently ask the team to prioritise issues into those that are important, those that are critical success factors (CSFs), and those that need to be addressed directly during the workshop.

Once the issues have been prioritised the value managers will facilitate a session

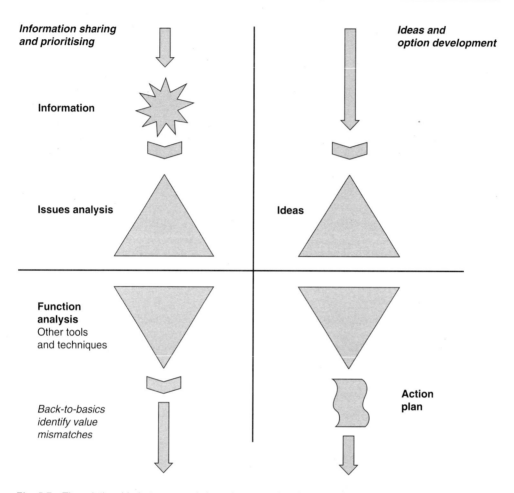

Fig. 5.7 The relationship between workshop phases and tools and techniques.

that explores the CSFs, focusing on those that have been identified as needing to be addressed during the workshop to determine what lies behind them. The process of doing this begins to challenge the team, moving them from stating opinions into the deeper areas of their belief, attitude and value systems. Functional analysis pushes a team even further; it constantly challenges assumptions and perceptions by asking *what* and more importantly *why* activities are being undertaken. FAST diagramming, if done correctly, is probably the most challenging technique. To complete the function logic diagram necessitates constantly asking why. Teams find this constant questioning and creating the logic tree very demanding. It is usually at this stage that value conflicts will emerge, with the FAST diagram attempting to resolve them through open and frank debate.

Figure 5.8 indicates these principles and the purpose of some tools and techniques.

	Information sharing and prioritising		
Issues analysis	Opinions	Beliefs	Innovate, generate ideas and option development
Issue prioritisation	Opinions and Beliefs		
Function analysis	Beliefs	Beliefs	Seeking consensus on the way forward
Element function analysis	Beliefs/Attitudes	Opinions	Action planning for implementation
Fast diagramming	Attitudes/Values		
	Back-to-basics identify value mismatches		

Fig. 5.8 Stages of the workshop, tools and techniques, and psychosocial dynamics.

An important requirement during the workshop process is to remain flexible. For example, the authors have found that there is a need to move back into value management during what has often appeared initially to be a value engineering study to improve designs due to the emergence of unresolved issues that should have been dealt with earlier in the project. Sometimes no one has previously questioned why a project existed and the authors have found a serious challenge to the fundamentals of the business project occurring during a value engineering study, when this type of concern should have been removed much earlier. This can be particularly acute during a charette style of study.

Sometimes an outcome of a VE study has been the cancellation of a project where a reversion to VM has shown an alternative strategic solution to the problem which did not involve construction. In one instance, due to the timing of a study (essentially a charette), the authors had to conduct a combined VM and VE on design development to date. The VM focused on unresolved strategic issues that emerged during the workshop process while the VE focused on design improvements on the existing solution.

Ideas tend to emerge in four key areas during the workshop process:

- Strategic improvements, usually in the domain for the client to solve.
- Design and construction improvements, usually in the domain for the technical delivery team to solve with inputs from the client.
- Project management improvements, usually involving improvements to value team working, project management processes and procedures or project structures.
- Risk, usually involving improvements to the management of risk.

It is not uncommon for separate working groups to be tackling each of these areas during the option development stage of the workshop process.

The workshop phase is also the arena in which ideas previously discarded can re-emerge and where innovation takes place, with good ideas often coming at the interface between disciplines. The dynamics of the workshop phase within a value study have also led to an observation that there is a 'right time' for an idea to emerge or re-emerge within an existing team, and the workshop process permits this to happen.

Workshop duration

The benchmarking study revealed a wide range of durations for the workshop phase and these have been highlighted in the section dealing with project study styles. However, a series of observations can be made concerning workshop duration.

There is no doubt that market economics is driving shorter workshop durations. However, a shorter workshop has to be balanced with the minimum time that a team will take to perform effectively and those offering value management services should be ready to advise clients when a VM/VE workshop is inappropriate. Workshop costs are significant and workshops should be used sparingly and only when necessary. There may be other ways of achieving the same results without resorting to workshops. At times, the authors have recommended not going into the workshop phase because the orientation and diagnostic phase has indicated this would not provide the best value solution. A facilitated or chaired meeting may be more appropriate in which case it should not be called VE/VM. However, if the meeting is called to resolve competing value systems, the domain of VM/VE is appropriate but this takes time and it may require sustained, intensive activity to achieve it. The dynamics of the workshop process discussed above attest to this.

The workshop phases are therefore about sorting out value systems, function analysis and introducing innovation into projects, processes, services and systems. It is not a box-ticking exercise or a surrogate facilitated meeting conducted in a workshop format and badged as VM/VE. The choice of workshop duration should revolve around:

(1) The extent of value system constellation required. If value systems are in alignment then shorter duration workshops are possible. Where value systems are misaligned, need re-aligning or the value thread is in danger of breaking or may have broken, longer duration workshops are required.
(2) The degree of perceptual and behavioural change required in a team.
(3) The extent to which there is a planned value programme in place. Where there is a structured value programme in existence shorter duration workshops are possible, where teams are well formed and the programme of activity well understood.
(4) The purpose of the workshop and its context within a value study and value programme.

(5) The extent of development of solutions required: is it to be undertaken during the workshop phase or external to it? The key consideration here is that, while a value team is together, more novel and innovative solutions can often emerge. If development is required as part of the workshop phase this would be unlikely during a shorter workshop.

It is the authors' experience, however, that research has resulted in efficiencies in workshop delivery and that team productivity can be enhanced with appropriate use of the tools and techniques described. Developments in the O&D stage mean that the workshop can start at a high rate of production and the use of tools such as the issues analysis can draw vast quantities of information in a relatively short time. Facilitation techniques mean that teams are close to 100% task orientated with little maintenance mode time. However, as stated above, the dangers of running a workshop on the rails laid down by the agenda are significant and a balance has to be struck.

Observations by the authors suggest that *one-day workshops* can work well provided the study is focused and the value problem is straightforward; there is a clear set of deliverables required from the workshop phase and this forms part of an ongoing process. Strategic studies work well as one-day workshops but only with good preparation during the orientation and diagnosis phase. Shorter duration workshops generally take the same overall study time as a longer duration workshop since the orientation and diagnosis phase will be extended to take account of increased preparation. Often considerable benefits in terms of team building and team focus can be achieved by starting a one-day workshop on the previous evening. The evening presents an opportunity for those involved to orientate towards the project and perhaps to commence the issues analysis. An evening meal gives the opportunity for the team to get to know one another and the fact that everyone is together at the same hotel ensures a prompt start on the morning of the workshop (provided extended team building does not continue into the early hours in the hotel bar).

Two-day workshops, with an evening before if possible, are optimal for most types of complex studies. Issues can be fully explored, with outline development taking place within the workshop process. They also provide sufficient time for value constellations to be aligned or re-aligned. Where the value managers have been unable to convince commissioning clients to opt for a longer duration workshop, a good compromise is an evening and full one-day workshop phase as described above.

Three-day workshops usually permit significant option development to take place within the workshop. However, the authors' experience suggests that these are now rare.

Team dynamics and managing teams during workshops

Managing team dynamics and planning team processes is an important skill of the value manager. The authors have found that skill sets are conditional on team size. Research[12] has indicated that team sizes of between six and ten people can work cohesively together. As team size goes beyond this range cliques begin to form and teams become less easy to manage. Teams of four to six can become very task focused and this is an optimum size for working groups where parallel working may be necessary. As mentioned previously, the requirement to remain flexible during the workshop phase comes to the fore in planning value team processes. The initial strategy for team working will have been worked out during the orientation and diagnosis phase of a study as the workshop agenda is being put together. Depending on the type of study, small value teams can be led by one facilitator for most of the study. However, if parallel group working is required the presence of two facilitators provides a distinct advantage in maintaining momentum, deciding workshop tactics and keeping small groups working on track. The authors expect most workshops will have parallel working group activity. Large value teams require the facilitator(s) to plan for parallel group working from the outset. The value manager's skill set becomes one of keeping the pace and momentum across working groups consistent; ensuring groups are working to a common agenda; deciding when to hold plenary sessions outside of those already planned in the agenda; and synthesising information quickly to ensure that workshop objectives are constantly kept to the fore.

Large teams – those in the order of 20 plus people – may require breaking down into smaller working groups from the outset. Plenary sessions will be planned at critical points in the agenda to bring working groups back together to present their findings and to refocus groups in order to achieve workshop outcomes. Plenary sessions are also a control and coordination mechanism within the workshop process for large teams. The location and physical resources available for large team working become important. A large room with plenty of wall space eases cross working group corroboration and coordination. Breakout rooms assist when periods of quiet and intense working are needed. Small value teams (six to twelve people) can often be retained as a single working unit for much of the workshop, although separate working groups may occur during the development component of the workshop. However, regardless of small or large team size, it may become obvious that as workshop activity evolves it becomes tactically necessary to break the whole team down into working groups or to reconfigure working groups as issues emerge and need to be resolved or the Workshop Phase dictates. Parallel group working is normal during the development stage of the workshop, regardless of team size.

The largest team that the authors have facilitated during the workshop process was 50. This was for a partnering workshop for a social housing scheme where there were many and diverse stakeholder groups, including supply chain members and those from the local communities. This required parallel group working from the outset and a team of four facilitators. At different times the working groups were operating on the same problem, on different problems or were working as one comprehensive team during plenary sessions. An important point was that the facilitators were familiar with with another's styles and approaches. The role of the lead facilitator included the additional task of ensuring that working groups were following the agenda, individual facilitators were keeping to time and task, and were integrating information as it emerged from workshop activity in the context of overall objectives to be achieved. This particular workshop also drove innovation in study process, methodology and techniques.

Team size has also caused the authors to think about different methods for functional analysis. Observation has indicated that using FAST diagramming is difficult with large teams. The function matrix technique has been developed for use in such situations. However, parallel group working to construct a number of FAST diagrams also has its merits since it can test for common understanding of a project across a large team. Parallel working groups can perform important problem solving activities during the workshop process. First, sub-groups can work on the same problem. This provides a degree of validation for robust outcomes through a much deeper analysis of the problem since more minds are brought to bear on the same problem. Second, sub-groups can work on different problems. This broadens the analysis and makes greater use of the full team resource.

The discussion on team dynamics reinforces the fact that the value manager as process manager is acting as a facilitator of team processes during the workshop phase. However, a series of uncertainties can confront the value manager as workshop facilitator:

- The value team may turn on the facilitator if they find the outcome of the process difficult to handle or that the use of the tools and techniques are highlighting issues they do not want to confront. It is important for the facilitator to remain detached from the dynamics of the situation and constantly to remind the team of the problem at hand and that they have to find a solution and a way forward.
- Professional jargon can act as a barrier to understanding where non-professional people are present. The facilitator needs to remind individuals to use language that everyone can understand.
- With politically sensitive studies it is essential that all key stakeholders are present to corroborate information. Drawing on information from interview transcripts is not a good substitute for having access to real time information from people dealing with the value problem on a day-to-day basis. It also improves the efficiency of the value process since it builds commitment to outcomes.
- The presence of a project sponsor or client representative able to take executive decisions at the workshop is essential since the workshop process may uncover things that need immediate decisions. It maintains momentum and permits the value team to find solutions to difficult problems as they arise. The presence of an individual holding this role also keeps the team focused on client issues and offers some control over team dynamics.
- With politically sensitive projects or value problems, the team may grow due

134 Method and Practice

to this very reason: nobody wants to be left out. The issue becomes one of being able to handle large teams or controlling numbers.

Facilitation

Research evidence on leadership highlights that leaders should adapt their style to the situation. The same argument applies with facilitation. Figure 5.9 indicates facilitation styles that have been observed in practice.

Seeks consensus	**Autocratic**
• Avoids conflict at all costs • Always seeks common ground and minimises differences	• Steers and leads the team from the front • Constantly intervenes • Dominates team dynamics
Laissez-faire	**Guiding**
• Lets the team go its own way • Social interaction is more important	• Chooses the appropriate style to suit the situation • The 'conductor' of team dynamics • Manages the 'process'

Fig. 5.9 Facilitation styles.

The figure is self-explanatory in terms of each of the styles. However, the argument proposed here is that a facilitator should be able to operate in all four quadrants as the situation dictates, and not permit one style to dominate their approach. This suggests that facilitation style has a degree of choice associated with it. Another issue that often occurs during the workshop phase is *leader transference*. It is a particularly difficult problem to handle and raises ethical issues for the value manager as facilitator during the workshop process. The phenomenon can become particularly acute when a workshop team is facing a highly ambiguous situation and where perhaps there is no obvious clear way forward or where apprehensions are high.

Value management teams will normally have a combination of leaders present, both formal and informal[13]. Leader transference occurs when the value management team hands over leadership and full responsibility for finding a way forward and solving their value problem to the workshop facilitator(s). The process can occur subtly. It is evidenced, however, when the value team starts to use phrases to the facilitator encapsulating the term 'you', rather than 'we' or 'I', need to find a solution. Strong evidence for transference is when the value team is consistently looking to the facilitator to find them a way forward and the psychological pressure moves from the team to the facilitator. The phenomenon can happen at any stage in the workshop process, within a matter of minutes of the workshop commencing or at a time well into it.

It is important to be aware of the issue since the danger is that the facilitator moves away from being process manager into becoming problem solver and member of the team. This is not their function. However, using the phenomenon once it has occurred becomes part of the 'tools and techniques' of the value manager as facilitator during the workshop phase to control the process of the workshop while recognising the responsibility of having gained the confidence, trust and authority of the value team. The best and only solution to the phenomenon is to remind the team gently and consistently that it is *their* responsibility to find solutions and an appropriate and agreed way forward, not the value manager's.

A further difficulty that can sometimes occur because of team dynamics is when the facilitator is used as a sounding board during tea, coffee and lunch breaks. This is likely to become particularly acute with sensitive or politically difficult value problems. Again, this raises ethical issues for the facilitator. Differing views may be presented to the facilitator on what is required, what is the right solution or what can or cannot work. The position taken here is that this is part of further information gathering, and it is not the role of the facilitator as process manager to take any one point of view as being valid. Rather, these views need to be aired publicly as part of the workshop process. If the individual concerned is not prepared to discuss them publicly, the facilitator should enquire if they are to be discussed in an open arena and then find a way of doing so. Otherwise the issue should be left as part of information that may or may not come out as part of the workshop process.

Tools and techniques in the workshop phase

The benchmarking study identified tools and techniques used during the value management process which are described in detail in the Toolbox. A number of observations can be made about their use during the workshop process. The main purpose of tools and techniques during the workshop phase is to elicit, structure, restructure and present information to attendees. They form the toolkit of the value manager as facilitator and include:

- Opening-up techniques: they open up information for further exploration, an example is an Issues Analysis.
- Closing-down techniques: they draw on existing information funnel its use down for a specific purpose, an example is a FAST diagram.
- Deepening understanding: functional analysis and FAST diagrams are prime examples.
- Eliciting information: operating at a surface level, an issues analysis is a quick technique to use for eliciting information, structuring it and presenting it visually.
- Structuring and restructuring information: FAST diagrams and decision matrixes are examples.

The important things for the value manager as facilitator to know are the strengths and weaknesses of techniques and how they can be used singly or in combination to commence, improve or reshape value team dynamics in order to improve information sharing and synthesis and to permit options to be generated and decisions made.

Presentations are a useful technique for either opening a debate for further work or for closing a topic. At the commencement of a workshop they are normally used to open a topic and at the final workshop presentation to close topics. During the workshop they can be used in either way. The facilitator must be aware of the intention of the presentation before using it.

Issues analysis is an opening up, structuring and restructuring technique normally used at the commencement of the workshop to elicit information from across the whole team. It is used to share knowledge, identify gaps in the team's knowledge or identify areas that the team may feel uncomfortable discussing openly at the outset of a workshop. An issues analysis can capture learning from previous or similar projects, capture current as well as future issues that need to be considered and identify areas of risk for further analysis. It is also used to commence team building, getting a value team relaxed with itself, and can be adapted either for whole team or parallel working group activity. The authors use it on nearly all occasions, but exactly when to use if can be a matter of either workshop strategy or tactics. For example, on politically sensitive projects or where a team is apprehensive about the process, the authors would tend to use it early in the workshop to get a value team talking to each other and relaxed. However, where time is short the authors may start the workshop with functional analysis and use this to generate a much more focused issues analysis. Due to the large amount of information that can be generated on sticky notes during an issues analysis session, the authors have adopted a prioritisation technique of using coloured dots.

The full issues analysis, with dotting for prioritisation, is fast, efficient, dynamic and visual. On completion, the team is fully aware of what it needs to do and where it needs to go. It remains as a constant reminder for the team throughout the workshop process. Often many functions for subsequent functional analysis will be found on the issues analysis. The dotted items can also be used for subsequent small group working and many ideas and solutions may have already been identified as part of the process. As with any voting technique the dotting technique can be biased because of team composition but the important point is that the dotting technique is used to identify important areas of information. If a significant part of the value team have all identified a certain issue as important then that is an essential piece of information.

Drivers analysis is a powerful technique and can be utilised in different forms at project, service or design level. It forces team to think about what is driving a project or design. The drivers analysis leads towards capturing and encapsulating value drivers which can subsequently be used to refine functional analysis and elements and components function diagramming.

Function analysis in some form is always used by the authors in value studies. It is a powerful technique for constantly asking *why* and *what*. The authors have found it useful for identifying the strategic purpose of a project (a strategic FAST diagram). When used in its matrix form it defines both the business project in terms of strategic needs and wants and the technical project (the physical product) in terms of its functional needs and wants. The function matrix diagram if used first can be translated subsequently into a strategic FAST diagram and a technical VE element and component diagram. The authors have also used SMART[14] as input into the development of a brief. However, one observation that has come from constructing function diagrams of different forms is that it they are only as good as the team present in the workshop. Different teams will develop different function diagrams since these are statements of value criteria or value systems. This reinforces the need to ensure that the right team is present during the workshop phase. At a basic level functional analysis is using natural language to structure a problem and order value constructs[15].

Functional analysis and function logic diagramming are powerful techniques during the workshop process. They can:

- Take a team back to basics.
- Identify the real 'value problem'.
- Identify why to invest in this project at this point in time and no other project.
- Confirm the need for architectural features.
- Determine the purpose of a department, team, or division within an organisation, or even an organisation itself, for example, within a supply chain.
- Bridge the interface between different teams working on a project to test for common understanding.
- Generate ideas using every function and hence generate large quantities of ideas.
- Reframe the problem and challenge fundamentally a team's assumptions and current logic.
- Create a value focus: that is, are you spending your money appropriately?
- Be used to understand a client's perspective when not present by searching within information generated by the client for value and functional statements.

The authors have increasingly questioned the role of functional techniques at different stages of a project and for different purposes. Element and component functional analysis is appropriate for technically orientated VE studies. Strategic FAST and SMART diagramming is more appropriate for project strategy levels of analysis – concept and briefing studies. Equally, the authors have found that, depending on when a study is conducted and the constraints present, FAST diagrams can generate hostility in a team by taking them back to basics and questioning fundamentals. However, where it has been conducted in such circumstances, and its purpose in the study explained to the value team, it has usually been a necessary technique to adopt.

Over time the authors have also questioned the use of the 'why/how' logic on a

FAST diagram. The 'why' direction is powerful, regardless of whether FAST diagramming is being used strategically or technically. However, by its nature the 'how' direction forces a team in a technical direction when that type of analysis may not be required. Teams can become frustrated with this mechanical approach to diagramming when the logic does not intuitively stack up. The authors have amended the FAST diagramming technique to remove the 'why/how' requirement, especially on strategic studies. The 'why' analysis remains. However, a 'levels' hierarchy is subsequently used to order the diagram, with level 1 to the left of the diagram being strategic and subsequent levels (levels 2, 3, 4, etc.) moving to the right becoming more tactical. Equally, purists would argue that the function logic diagram should be completed once commenced. However, at times the authors have found that a value team has found it difficult to prioritise a diagram correctly until it has understood the problem in much more depth through iterative analysis, with the final structure emerging towards the end of a workshop.

Functional space analysis, user flow and spatial adjacency diagramming is an extension of functional analysis used on building projects. These techniques can be used dynamically in a workshop setting to assist a team either in understanding how an organisation intends to use space or in auditing designs in terms of space efficiency. Both techniques permit a functional analysis of space to take place and when used in combination with a SWOT analysis on a design provide a series of powerful tools for exploring value in design. User flow diagramming has also been used successfully on road and rail projects to guide a value team into creating a common visualisation of a route and subsequently identifying or clarifying major issues along the route.

Element function analysis is a further powerful technique that has been found to move the workshop phase forward when positions within the team have become entrenched. In this context comparative project benchmarking has been used in three main forms:

(1) A straight comparison of cost/m^2 and element cost/m^2 for similar projects.
(2) Histograms of the cost profiles of similar projects to explore the reasons for high and low cost elements within the project under study. This is a visual technique and is usually a precursor to identifying value mismatches and generating options and ideas for improvement. It is powerful for challenging elemental costs.
(3) The Building Cost Information Service standard element list is used to describe in outline the typical specifications that can be bought for a particular cost/m^2 on each element measured against the expectations of the value team. The process builds a holistic picture of the specification that may be purchased for a certain budget. This is a particularly useful technique where expectations may be out of synchronisation with budget constraints.

Risks have always been identified by the authors as part of the value process, although no formal risk analysis is undertaken within the workshop process. At one stage value and risk were viewed as different activities, with a value management study and a risk management study being undertaken at different times. However, increasingly the authors have combined value and risk analysis, viewing them as two sides of the same coin, with value opportunities and improvements being on one side of the coin and risk management on the other. Observation has indicated that each has and requires different mindsets. Value improvements are seen as seeking opportunities to improve a project and hence viewed in a positive light, whereas risk tends to be seen as threat focused and more negative in character. Given that they have different psychological mindsets associated with them it is far easier to conduct a value opportunity analysis first, followed by a risk analysis of value opportunities and then a full risk analysis. It is more difficult to move a team from being threat focused to value opportunity focused without a time period for moving mindsets taking place. Risks can be identified from the issues analysis either by using separate risk identifiers or exploring what lies behind important issues in terms of value opportunities and/or risks. Risks can be logged for further detailed analysis after the value workshop has been completed.

Summary of the workshop phase

To summarise, the workshop phase is where different value systems are brought together to improve a project, process, service or organisational function. The benchmarking study has identified the role of the facilitator during the workshop stage as critical to the success of a value study. The previous discussion has highlighted the fact that part of the skill set of a value manager is associated with deciding on the strategy and tactics of the whole study, including the strategy of the workshop phase within the overall study. The skill set of the value manager as facilitator of the workshop process is understanding team dynamics and interpersonal relationships and having a set of tools and techniques for eliciting, structuring and restructuring information to assist decision making and solve a value problem. During the workshop phase the value manager requires tactical skills to act as a process manager of a value team workshop.

The implementation phase

The benchmarking study revealed that the implementation phase is the one area where value management can easily fall down unless planned for as part of the process. The authors always address implementation as part of the orientation and diagnostic phase. It continues to be addressed at the end of the workshop phase with the development of an action plan, which draws the workshop together and allocates responsibilities within a specified period.

Review workshops or implementation meetings are good for embedding the implementation phase into a value study. They will normally be of half or one day duration and are best planned to occur two to three weeks after a value study took place. Review workshops are good for focusing the value team's thinking on implementation after a study has been completed and the report issued. Working groups and individuals report on progress since the value study has taken place. Review workshops also provide further focus for implementation via an updated action plan. Again, while desirable, they are rare in the authors' experience. The most common mechanism for the implementation phase is a meeting to review the final report and action plan with the commissioning client.

While the implementation phase has been identified as the Achilles heel of value management, the benchmarking study also revealed that some clients are very systematic in following through on the value generated and implemented ideas formulated during the workshop process. This requires systematic collation of workshop outcomes and tracking ideas through to incorporation.

The next section draws together the lessons from this chapter to redefine the value process and value studies.

5.6 Value studies: a revised process

The authors have drawn together in this chapter study styles identified in the benchmarking study, others from action research and observations from over 200 different types of studies. Unfortunately, in most instances value management has been used reactively to sort out situations once things have become difficult. However, the strength of VM can be harnessed if it is used proactively to forward think projects and programmes, designing new services and anticipating difficulties and opportunities in organisations facing change and restructuring. Where we have been able to do this the process has proved very successful. One of the important enablers for the whole process, which includes using it proactively, is support from senior management and from within value teams involved in the process. Equally, the obverse provides barriers to the process.

Earlier sections have described the value process and lessons learnt to date, including presenting alternative study styles. Over time the authors have reviewed and revised their own thinking and approach to value management. However, regardless of the study style adopted the important reference point is that value management is not just about running and facilitating workshops. This is only one phase in a more comprehensive and structured study process. The whole value process is about understanding, bringing together and integrating value systems to introduce improvements. Within projects it has to be placed within the project value chain framework described in detail in Chapter 7.

The benchmarking study of value management identified a number of critical success factors:

(1) The use of a multi-disciplinary team with appropriate skill mix.
(2) The skill of the facilitator.
(3) The structured approach through the VM process.
(4) A degree of VM knowledge on the part of the participants.

(5) The presence of decision takers in the workshop.
(6) Participant ownership of the VM process output.
(7) Preparation prior to the VM workshop.
(8) The use of function analysis.
(9) Participant and senior management support for VM.
(10) A plan for implementation of the workshop outcomes.

The authors' view now is that, while these CSFs remain relevant, the discipline of value management has moved on, not least in its relationship with risk and risk management. Alternative study styles have been presented which demonstrate the flexibility of the process as a value focused group decision support system. An additional CSF can now be added to the above which includes designing, and delivering an appropriately structured value process tailored strategically and tactically to suit the value problem under study.

The change in emphasis of value management over time, which includes pressure from many clients to reduce workshop time, has also meant that the authors have sought process efficiencies, some of which have been described above. However, it is important to emphasise that the power and strength of the process will be removed if VM becomes reduced to at best a procedure that must be undertaken or at worst a box ticking exercise. The authors are now advocating a much more tailored approach to VM. It has been argued by some that it may require renaming or rebadging to remove associations with a purely facilitated workshop activity. Value management is more than this; it is a process for structuring the interaction of often disparate value systems to find an acceptable outcome to move a project, service or organisation forward.

The need to find mechanisms that structure value systems and maintain the value thread intact throughout, for example the project value chain, is consistent with the findings of Wallace[16] during the design process. He found that multi-disciplinary coalitions align and re-align as the design process progresses regardless of contract type, project complexity and procurement route. Value management provides a mechanism that through a structured, challenging, analytical and mediated process permits value systems to coalesce to the benefit of the client. By bringing the right team together at the right time, value management focuses on value system evolution and resolution. However, this makes it a change orientated methodology.

When used proactively value management is about aligning the project value to the value systems of the client from the outset. Observation has indicated that on projects there are 'hard' and 'soft' gates. The former cost significant amounts of money to correct once gone through while the latter present opportunities to retrieve a situation. When value management is used reactively is usually involves re-aligning value systems that have become distorted for whatever reason or attempts to re-assemble a value thread that has become strained or broken. In many instances, once the value thread is broken, for example where a project team becomes dysfunctional, it can become an uphill struggle to re-assemble it, or the study may prove this is not be achievable.

142 Method and Practice

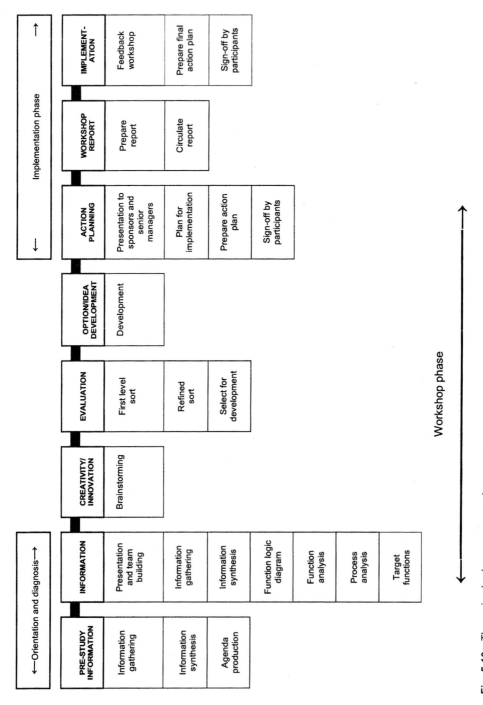

Fig. 5.10 The revised value management process.
Source: Adapted from Male *et al.* (1998) *The Value Management Benchmark*[17].

Figure 5.10 presents the revised value management process, including some important activities that take place during each of the phases.

Deciding on the appropriate methodology can augment the process presented in Figure 5.10, namely:

- Deciding on the study style.
- Choosing the approach to the value management team:
    - Single value manager.
    - Two value managers.
    - A combination of the above, including a workshop recorder or other variants.
    - Use of the existing team of record. The advantage of using the existing team is that they will have in depth knowledge of the value problem under study, costs are reduced and the implementation rate for improvements generated from the study will be higher. However, the existing team may have entrenched ideas and may bring few new ideas to the process. The benchmarking study indicated this was the preferred method in the UK and Australia and in the private sector in the USA. The role of the value manager as facilitator and process manager during the workshop phase is particularly important with this approach. The focus will be on challenging existing team thinking.
    - Use an independent team of experts tailored to suit the value problem. This may or may not include client representation. The choice of the right team is essential. The strengths of this approach are that the team will bring more fresh ideas to the process and convey technology transfer to the value problem, will have no preconceived ideas about the value problem and, if chosen appropriately, should produce a good team dynamic. However, costs are considerably higher, there will be resistance to ideas generated by the independent team from the team of record, the learning curve is steeper and the study is likely to take longer. The lead value manager will be more focused on process management and less on challenging the independent value team's thinking. The authors have used this approach, with and without client representation present, for project audits and facilities programming studies.

This chapter and preceding chapters have provided the theory, approaches and observations of the authors on value management in practice. Subsequent chapters will expand on the arguments presented here.

5.7 References

1. Male, S., Kelly J., Fernie, S., Gronqvist, M. & Bowles, G. (1998) *The Value Management Benchmark: Research Results of an International*

	Benchmarking Study. Published Report for the EPSRC IMI Contract. Thomas Telford, London.
2.	McNiff, J. (2000). *Action Research in Organisations*. Routledge, London.
3.	Glaser, B. G. & Strauss, A. L. (1967). *The Discovery of Grounded Theory: strategies for qualitative research*. Weidenfeld & Nicolson, London (printed in USA).
4.	Locke, K. (2001) *Grounded Theory in Management Research*. Sage Publications, London.
5, 6, 7.	Male, S., Kelly, J., Fernie, S., Gronqvist, M. & Bowles, G. (1998) *The Value Management Benchmark: A Good Practice Framework for Clients and Practitioners*. Published Report for the EPSRC IMI Contract. Thomas Telford, London.
8.	Kelly, J. R. & Male, S. P. (1995) Facilities programming. *Proceedings of Construction and Building Research Conference (COBRA)*, Royal Institution of Chartered Surveyors, Edinburgh, September 1995.
9.	Kelly, J. R. & Male, S. P. (1993) *Value Management in Design and Construction: The Economic Management of Projects*. E. & F. N. Spon, London.
10.	Blyth, A. & Worthington, J. (2001) *Managing the Brief for Better Design*. Spon Press, London.
11.	Miles, L. (1972) *Techniques of Value Analysis and Engineering*, 2nd edn. McGraw-Hill, New York.
12.	Hunt, J. W. (1992) *Managing People at Work: a manager's guide to behaviour in organizations*, 3rd edn. McGraw-Hill, London.
13.	Kelly, J. R. & Male, S. P. (1993) *Value Management in Design and Construction: The Economic Management of Projects*. E. & F. N. Spon, London.
14.	Green, S. D. (1992) *A SMART Methodology for Value Management*. Occasional Paper No. 53. Chartered Institute of Building, Ascot, Berkshire.
15.	Kelly, G. (1955) *The Psychology of Personal Constructs*, Vols 1 & 2. Norton, New York.
16.	Wallace, W. A. (1987) *The influence of design team communication content upon the architectural decision-making process in the pre-contract design stages*. PhD thesis, Heriot-Watt University.
17.	Male, S., Kelly, J., Fernie, S., Gronqvist, M. & Bowles, G. (1998) *The Value Management Benchmark: A Good Practice Framework for Clients and Practitioners*. Published Report for the EPSRC IMI Contract. Thomas Telford, London.

Part 2 Frameworks of Value

Chapters 6, 7 and 8 outline the context within which the frameworks of value reside. These chapters bring together and synthesise the theories of value, relate value to quality systems and demonstrate a method for the evolving of the client's value system. The authors intend this section to be theoretical, thought provoking and challenging to those who are currently engaged in the research and practice of value. Reference is made to the work of four PhD graduates whose highly successful research explored the depth of value in its various forms.

Chapter 6 sets up the parameters for value management. It discusses 'value' and 'value systems', and defines value management and value engineering as used by the authors. Value management, as a discipline, brings together ideas, concepts, models, tools, and techniques from economics and finance as well as organisational behaviour and social psychology. It also integrates strategic management and project management disciplines. The chapter argues that value management is business project focused while value engineering, as a subset of value management, is more technical project focused. The chapter sets out a typology of clients to introduce the idea of different client value systems, each of which must be understood in order to tailor a value study. The chapter concludes by raising some ethical issues within value management since it is a process of resolving and reconciling different and diverse value systems.

Chapter 7 builds on Chapter 6, drawing together research work on the strategic phase of projects and the project value chain in the context of corporate and business value and their application to a two-stage briefing process for projects. The chapter also describes an action research programme of studies that has been underway since the early 1990s involving a wide range of projects undertaken using a variety of procurement systems. The issue of projects and the individual's place within the project at each successive stage of the project is analysed and characterised as a relay team where each individual has to pass on the baton to the next person smoothly and efficiently to stand a chance of winning the race. The same principle operates within the project value system. Participants at each stage of the project life cycle have to pass on the 'baton' efficiently and effectively to those involved in the next stage. This creates the structure to the project value chain.

Chapter 8 discusses the relationship between cost and quality, exploring the place of structured value criteria within an existing quality environment. It discusses the concept of quality and demonstrates that value management and value engineering are powerful methodologies when total quality management and quality assurance have exposed a project to change and/or to improve procedure. An example of best value is used to illustrate this relationship. The chapter concludes by bringing together information from the chapters in Part 2 to define the client value system using a structured technique.

6 Value Context

6.1 Introduction

This chapter sets up the parameters for value management. It discusses 'value' and 'value systems', and defines value management and value engineering as used by the authors. Value management, as a discipline, brings together ideas, concepts, models, tools, and techniques from economics and finance as well as organisational behaviour and social psychology. It also integrates strategic management and project management disciplines. The chapter argues that value management is business project focused, while value engineering, as a subset of value management, is more technical project focused. The chapter sets out a typology of clients to introduce the idea of different client value systems, each of which must be understood in order to tailor a value study. The chapter concludes by raising some ethical issues within value management since it is a process of resolving and reconciling different and diverse value systems.

6.2 Defining value

This section looks at different interpretations of value, drawing on ideas from economics, psychology, and social psychology.

Bell[1] discussed in some detail the concept of value. She noted that historically the concept has been influenced heavily from an economic perspective and is normally expressed as the ratio of costs to benefits. The primary mechanism for communicating value decisions has been in monetary terms. Other authors have discussed value in terms of, for example:

- Use qualities.
- Esteem features related to ownership characteristics.
- Exchange properties related to the market place.
- Cost characteristics such as the sum of labour, materials and other costs, including the cost of finance.

Value, as a term, can also involve perspectives from a producer, customer, consumer or user. From a value perspective, these can be considered as different generic roles in understanding 'value', for example:

(1) The producer provides at a price a physical object or service for consumption.
(2) A customer buys the physical object or service at a price from the producer as part of an exchange relationship.
(3) The consumer utilises the product or service for a purpose, which may involve use.
(4) The user is looking to use the product or service on the basis that it has appropriate functional aspects contained within it that will provide satisfaction and benefits.

Value can also encompass a process, product or service. Normally, these generic roles overlap when individuals buy products or services for household use. However, with organisations varying in their levels of complexity these roles may become dis-aggregated and diffused throughout the organisation as part of its structure.

As mentioned above, utility and the satisfaction obtained from use also enter the equation when discussing value. Value can be defined as the intrinsic property to satisfy[2]. Satisfaction has a psychological dimension to it and can relate to decisions made by an individual or a group. The moment more than one individual is involved in decisions about value and satisfaction, complexity of decision making comes into play. Equally, individuals or groups of individuals may sit in different departments, organisational units, organisations or firms with different ownership characteristics. Value decisions then become very dependent on:

- The complexity of perceptions involved.
- The context within which judgements about value and satisfaction are made.
- The number of interfaces that exist between individuals, groups of individuals, organisational units, organisations and firms that decide on value and the benefits and satisfaction derived therefrom.
- Power of different individuals or organisational units.

The social and psychological dimension to value has an important part to play in making choices and managing projects and organisations. *Values* comprise part of the cognitive structure of an individual; that is, the manner in which he or she sees, perceives and thinks about the world, structures information, takes action and behaves. Values, attitudes and beliefs have emerged from social psychology to provide a framework for explaining the way people behave in social situations, perceive others and change and adapt their behaviour over time. Gross views *attitudes* as an integration of beliefs and values[3], whereas *beliefs* comprise the knowledge and information we have about the world. Attitudes involve three classes of response: those involving feeling and emotion; those involving perceptions, and concrete actions, either intended or real[4]. Attitudes provide shortcuts for relating to and interpreting events, situations and objects. Values result in judgements about what is good, bad, desirable, etc. Gross, on reviewing the research evidence, concludes that individuals will have thousands of beliefs, and hundreds of attitudes but only a few dozen values[5]. Values operate at a much deeper level of cognitive functioning than either attitudes or beliefs.

Yin[6], in a seminal work on the study of human communication, notes that attitudes are usually tapped through verbal testimony, i.e. *opinions*, although not without difficulty[7]. Allport *et al.*[8] provide the best known classification of values, using a measuring scale to assess an individual's value orientations. They identified six distinct value orientations which can be viewed as 'ideal' types:

- *The theoretical*: This value orientation is focused on the discovery of truth. The individual is concerned with rationality, critical appraisal, problem solving and the empirical, with a view to ordering and systematising their knowledge. The theoretical person is likely to reject notions of beauty or utility. Their focus is on knowing how things work and the underlying principles.
- *The economic*: This individual places a high emphasis on usefulness. They are practical, business focused and with an interest in making money. Knowledge exists to be applied. Their focus is money and finance.
- *The aesthetic*: This individual places highest value on form and harmony and sees truth as beauty rather than seeking truth for its own sake, as does the theoretician. The aesthetic is interested in social welfare but has a tendency to be individualistic and self-sufficient. They tend to be at odds with economic values. Theoretical, economic and aesthetic individuals are seen as cold and inhuman.
- *The social*: The social individual tends towards the philanthropic and altruistic and is kind, sympathetic, unselfish and concerned for the welfare of others. There can be a close affinity with the religious value.
- *The political*: This individual is interested in power, yearning for personal power, influence and renown above all else. Their focus is on political systems (although not necessarily governmental politics) and power structures. Leaders with a power dimension to their personality would fall within this domain.
- *The religious*: This individual is concerned with unity, the mystical, morality and a higher meaning in the cosmos. They seek to understand and experience the world as a unified whole.

An individual demonstrates a range of these characteristics but has one overriding orientation.

Attitudes, beliefs and values are notoriously difficult to change because they are so deeply held. Social psychologists argue that individuals seek consistency between their thoughts, beliefs, values and attitudes, attempting to appear rational to others as well as to themselves. They feel psychological discomfort when these are inconsistent. This often occurs when an individual has made a difficult choice or decision or is experiencing hardship, making sacrifices that turn out to be pointless or becoming involved in behaviour that is inconsistent with internal attitudes and beliefs. When a person feels strong internal inconsistencies this will lead to attitude change[9].

To summarise, opinions operate at the surface of awareness, can represent attitudes that are at a much deeper level, and are usually tapped through verbal statements. Beliefs, representing an individual's knowledge of the world, cluster around attitudes, with values providing the judgemental component to attitudes. Attitudes, beliefs and values lie at the core of personality and are deeply held convictions and orientations about who and what an individual is, their place in the world and how they interact with others. They influence choices. Drawing together work from Yin and Gross, a useful model for thinking about values in a social and psychological context can be constructed . The model is set out in Fig. 6.1. The circles are deliberately indicated as dotted in certain instances to demonstrate a permeable boundary between psychological layers that interact and overlap but are conceptually different. The stress on exploring the social and psychological dimensions to value provides a framework for thinking about the processes going on within a value management or value engineering study.

In the philosophies of ancient Greece value was seen as a property of goods or services. The Greeks believed that there were certain primary or essential principles in our environment which gave value to the items they inhabited; thus ethics contained 'the good' religion 'the holy' and aesthetics 'the beautiful'. When the indwelling principle was present the object had value and when it was absent the object was worthless. Shillito & De Marle[10] offer the view that value is the primary force, a potential energy field that motivates human actions in respect of the objects they desire or need. Therefore the value of products or services can be represented by the presence of positive and negative features that attract or repel the customer. Understanding these gives the value system of the customer. In a similar manner, Mudge[11] quotes Aristotle who makes seven classifications of value being economic, moral, aesthetic, social, political, religious and judicial.

A number of value engineering texts state that value denotes a relationship between function, cost and worth[12], or function, cost and quality where value equals function plus quality divided by costs[13]. Seeking functionally equivalent options to meet users needs, desires and expectations, with varied cost, quality and worth attributes, is undertaken usually for the purpose of improving value. This concept can be summarised as 'the lowest cost to reliably provide the required functions or service at the desired time and place and with the essential quality'[14]. Value is expressed in these texts as:

- Exchange value, which relates to worth or the monetary sums for which products can be traded. This may be different from market value defined as the sale price of a product under the voluntary conditions of a willing buyer and a willing seller.

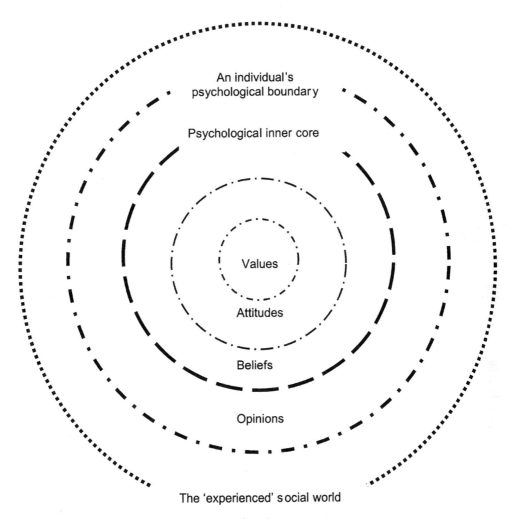

Fig. 6.1 A social and psychological framework for value.

- Esteem value, which relates to oneself or the monetary amount to be paid for functions of prestige, appearance and/or other non-quantifiable benefits. It may also refer to the monetary measure of the functions of a product that contribute its desirability for sale (including obtaining for the sake of possession).
- Use value, which relates to need or the life cycle cost considering user function only or those properties that accomplish a use by some work or service.

Fallon[15] states that the concept of value is encapsulated in the phrase 'an idea that relates objectives to one another and the cost of obtaining them'. Fallon also quotes from Adam Smith's *Wealth of Nations* on the paradox that extremely useful goods, such as water, have little or no exchange value whereas certain other goods of comparatively little use, such as diamonds, have great exchange value. Miles[16] states that the degree of value in any product depends on the effectiveness with which every usable idea, process, material and approach to the problem have been identified, studied, and utilised. Value is therefore considered good if a product of lower or equivalent cost to its competitor contains a somewhat better combination of ideas, processes, materials and functions. Miles is also quoted by Fallon as having stated that poor value is a people problem.

Financial management texts introduce the concept of stakeholder value. Scott[17] states that the principal goal of management is to maximise the level of sustainable growth in profitability and thereby enhance shareholder value, defined as the maximising of returns to those who have an ownership stake in the business. Porter[18] discusses value in the context of the economics of substitution, making the point that different buyers will evaluate substitution differently when faced with equal economic inducements for a functionally suitable equivalent product or service. The understanding of the buyer's value system in this situation is fundamental to the supplier's continued existence. Harvey[19] makes a similar point, stating that the value of a product or service is significant in economic terms only when a person is prepared to give up something in order to obtain it. An individual's value system will determine how much to give up and thus different goods will have differing values. Value is therefore measured in terms of the opportunity cost to the owner or purchaser. Finally, the European/British Standard on value management[20] defines value in the context of VM as the relationship between the satisfaction of need and the resources used in achieving that satisfaction.

To conclude, value is defined here as the equivalence of an item expressed in objective or subjective units of currency, effort, or exchange. Equally, value can be measured on a comparative scale that reflects the desire to obtain or retain an item/object. In this respect there are two components to value. One is an *objective* component and stems from looking at value from an economic perspective. The objective component of value can be defined by hard evidence. Examples include:

- Cost: a measure of the input resources used to create a physical component, object or service.
- Price: a measure of how the marketplace defines the functional benefits and satisfaction anticipated and expected by the customer, consumer or end user and provided potentially by the producer in an exchange relationship.

The *subjective* component for value is more difficult to define explicitly. It derives from individuals and groups making choices about costs and price, and the benefits and satisfaction derived or expected from consumption. Judgements about aesthetics, for example, fall into this domain. Opinions, attitudes, beliefs, social interactions in groups and teams, culture – both organisational and national

– and hierarchy influence the subjective component of value which is rooted in the psychological and social processes within and between individuals. Whether or not these objective or subjective measures can be evaluated in monetary terms, money is the unit used as a meaningful means of expression. The value in this context will relate to an individual or group acting within a single value system.

The next section develops these ideas further using concepts from systems thinking.

6.3 Value systems and clients to construction

Value systems

This section explores some of the key concepts in systems thinking that will set the parameters for a later discussion of value systems in general and client value systems in particular. A system, as a complex, organised and adaptive whole[21], comprises:

- Environments
- Boundaries
- Hierarchies
- A common purpose
- Objectives (that can often be in conflict)
- Interdependencies

Value systems are fundamental to value management. The above components provide some of the ideas that are important for defining value systems for use in VM. The arguments behind the concept of a value system and the client's value system are built up over the forthcoming sections and also encapsulate ideas highlighted in the next chapter where value and project processes are brought together. The authors have adopted, therefore, a systems perspective since it addresses problems holistically. Soft system thinking, focusing on processes, people, structure and organisation, starts from the premise that a problem situation is often unstructured. In an organisational context this can be a particularly useful mindset to understand the functioning of organisations, particular stages in a project, or a project or projects as a whole. At other times a holistic approach to the problem may be required but with a much more deterministic, structured and hard systems perspective, using 'hard evidence' such as cost.

In bringing together ideas from the previous sections and 'systems thinking', there is, first, a 'value problem' which is trying to be addressed and solved; that is, an event, situation or situations where judgements and decisions have to be made using 'hard' and 'soft' evidence. The term 'problem' has a neutral connotation, more akin to that encountered in mathematics where something needs to be solved. Second, value problems are embedded within a wider *value system*. The value system will comprise 'actors' having various levels of interaction, for example:

- Supra-systems such as the global economy, nation states, national governments.
- Industries.
- Private and public sector organisations.
- Divisions, departments and teams.
- Organisational roles.
- Individuals.

The value system comprises any of the preceding where judgements are made about allocating and using resources. These decisions and choices will be based on the benefits and satisfaction that are derived and accrue once choices about how to spend money have been made and then executed.

The next section extends the ideas of a value system to include organisations, clients and the concept of the value chain within construction.

Organisations, firms and clients: the nature of the client in the value process

Regular procuring clients to construction are generally complex organisations operating in either the private or public sector. Some have ceased to be state monopolies or quasi-monopolies and have transferred this characteristic into the private sector through privatisation; historically, the rail infrastructure industry is a prime example. Others operate in the state sector but apply or are attempting to apply business principles to public sector organisations; universities fall into this category. Some are large, some small, others compete in the global economy and have become world class organisations and others aspire to this. Many have decided to remain in the UK and compete nationally, regionally or locally. Hence clients to construction are a heterogeneous group of organisations facing different environments and with a diversity of reasons for existing, with different objectives, cultures and value systems. The nature of clients as organisations will be explored further below.

Organisations are social constructs. They have formal and informal structures whose primary purpose is to reduce the variability in human behaviour and help managers achieve a common purpose for the organisation and co-ordinate tasks across different types of labour. Power is also exercised within an organisation's structure and information flows take place between the organisation's different constituents. Within the organisational structure people structure their expectations around situationally determined roles which, in turn, are affected by the formal and informal structures of the organisation, its technical system and the individuals themselves performing that role. The technical system, and technology, probably have the most pervasive impact on an organisation[22].

Organisations have strategies. When looking at the term 'strategy', however, there is no succinct, agreed definition of what it is. It is an outcome of a strategic management process. Johnson & Scholes[23] separate the process into three distinct areas:

- Strategy formulation
- Strategic choice
- Strategic implementation

Strategy, as an outcome of this process, can be defined in terms of a means–ends relationship. The means are a set of rules to guide organisational decision-makers; ends are measurable objectives against which organisational performance can be quantified. Strategy is also seen as a cultural web, a set of collective beliefs shared by an organisation about the direction in which it is going[24]. Hence strategy can be either explicit and stated in documents or implicitly understood as consensus throughout the organisation. Some writers view strategy as a resource allocation process within a defined business scope. Others view it as a series of outcomes that affect the goals of an organisation and its stakeholders. Some have gone back to its roots in the military, defining both strategy and tactics. From this perspective strategy is seen as where to compete, for example in an industry, industry segment or geographic location. Tactics, on the other hand, are the organisational devices, the resources and expertise used to achieve a strategy, for example particular organisational functions such as marketing, or parts of the organisation such as particular teams. The military perspective sees both strategy and tactics as having a short-term and long-term perspective attached to them. This is in contrast to some writers who see strategy as delineating important decisions that have long-term ramifications for the firm. The literature on strategic management also distinguishes a hierarchy of strategies[25]:

- Corporate strategy is holistic and concerned with the activities of the whole company across its organisational configuration of different business units.
- Business strategy is concerned with competing in different markets or industries.
- Operational or functional strategy is concerned with the production process or functional departments within an organisation.

From an economist's perspective a firm is a social organisation that exists to make profit. This is but one of its potentially many objectives or goals. A firm can also be construed as a series of contracts and relationships, which add value to customers by delivering a quality product or service, and potentially offer something distinctive in the market place. Some economists argue that organisations exist because it is more cost effective to handle transactions within a firm than through numerous discrete, ongoing, market based transactions[26]. Public sector organisations are more allied to beauratic forms of organisation, where the formal structure dominates and rules and procedures guide organisational behaviour. They are likely to perform activities with a greater social dimension to them since they are extensions of the State at various levels of government.

Other terms allied to the strategic management process are *core business*, *core competencies* and *distinctive capabilities*. The core business of a firm is derived from its distinctive capabilities, the latter being created from three primary sources[27]:

- *Organisational architecture*. This is derived from the social and commercial relationships that the firm has and comprises the internal relationships between employees; external relationships with suppliers and customers; and networks of firms involved in related activities with the firm. Crucial elements to organisational architecture are the relational contracts founded on trust and expectations which define the moral and psychological contracts that the firm has. These may or may not be underpinned by a legal contract.
- *Reputation*. Reputation, as a strategic asset, has to convey meaningful information to customers. It is built through a process of continued success. It decays easily through poor performance. Reputation embodies the long-term experience of the firm and creates the likelihood of repeat business. Reputation is worthy of investment as part of the firm's value creation process provided the cost of maintaining and upgrading reputation is less than the premium that can be charged on goods and services as a result.
- *Innovation*. As a strategic asset on its own it is difficult to secure rewards and a profit from innovation. However, when combined with organisational architecture, in particular, it can implant a process of continuous innovation in a firm.

Organisations are also said to have core competencies which represent the collective learning of the firm across its skill, production and technology bases[28]. They provide value to the customer. Core competencies are likely to be implicit and can be identified through three tests. First, they provide potential access to a diversity of markets; second, they should contribute significantly to perceived customer benefits; and third, they should be difficult for competitors to imitate and are likely to be rare, complex and embedded in organisational knowledge and practice. Core competencies should become the essence of core products. However, Kay argues that the notion and operationalisation of core competencies is illusive and it is better to focus on distinctive capabilities, as identified above[29].

Strategy lies at the heart of developing and delivering value. In a value management context the concepts of strategy, core business, core competencies and value are inextricably linked within the client organisation, be it public or private, large or small. They are also of major concern when a client brings together firms to deliver a physical asset through a network of suppliers of services and products – the construction supply chain. They find their expression in the conception and delivery of projects as outcomes of the client's strategic management process.

The next section will explore the idea of different client types and hence introduce the notion of different client value systems that will affect the value management process.

Client types and client value systems

One of the important considerations for value management in construction is the impact that the client (or customer) has on the project process. Each client has distinct requirements and value systems, driven by their own organisational configurations, business and/or social needs for a project, the external environment to which they have to respond and the manner in which they approach and interface with the construction industry. Many of these influences have been outlined above. The client also commences the process of procurement, bringing together skills and expertise through a project process to deliver a completed product – a physical asset of some type – to meet a business or social need, or both.

A number of distinguishing characteristics can be applied to clients[30], and can also provide clues about their different value systems. As mentioned earlier, clients separate into *public* or *private* sector clients. Public and private sector clients have different value drivers. Public accountability and more recently the Government's Best Value initiative are among some of the drivers for public sector clients as representatives of the State. They also have to take account of influences at European Community level. Private sector clients are much more heterogeneous. Influences can range from the impact of shareholder value, to time-to-market considerations and ownership considerations due to plc, private limited company or perhaps family business status.

Clients also differ in their level of knowledge of the industry. They may be:

(1) *Knowledgeable.* These clients will normally have a very structured approach in dealing with the industry and project delivery. Often this will be set out and summarised in a project manual, a set of procedures or guidelines. They will treat the construction supply chain and its members as 'technicians' to deliver a project or projects to meet their business need and/or social need. Internal or external project managers will act on their behalf as the interface with the construction industry. They will tend to be innovative with procurement methods and will generally be the volume procurers of construction services. They will place considerable demands on members of the construction industry and expect it to respond accordingly.

(2) *Less knowledgeable.* These clients will often have limited or minimal in-house expertise and knowledge of the operations of the construction industry. They will have partial or no appreciation at all of the complexities of construction. Evidence suggests that, depending on their initial contact point within the industry, they will tend to be directed into a traditional procurement path. However, this type of client may consult more knowledgeable clients depending on its business network.

Clients or customers to construction can also be classified as:

- *Large owner/occupiers*, who will use physical assets to support their ongoing strategic plan in meeting a business or social need.
- *Small owner/occupiers*, who will often react to change and approach the industry because their existing facilities are inadequate in some way.
- *Developers*, who view facilities as a method of making profit. They will trade the asset to achieve this, or see it as an investment to generate profit and look for business opportunities and available sites to ensure a quantifiable return.

A fourth dimension to client characteristics is the economic demand placed into the industry in terms of volume, its frequency and regularity, coupled with the extent to which standardisation may exist from project to project in terms of parts, processes and design[31].

- *Unique* construction occurs due to the distinctiveness of technical content, the level of innovation required or the extent to which the client requirement is for leading-edge projects that push the industry's skills and knowledge to the limit. With this type of project there is limited, if any scope, for efficiencies in process or standardisation and repetition[32].
- *Off-the-peg* construction, which tends towards the unique type above but where the possibility exists for standardisation, perhaps through repeat designs where clients may only require one or two buildings. The term *customised* reflects a better description since a design for a particular client is likely to have some foundation in previous designs undertaken for other clients but with some adaptation[33]. Some commercial offices and warehousing fall into this category.
- *Process* construction, which can occur when a client has repeat demands for projects and a high degree of standardisation is possible due to the volume placed into the industry. Efficiencies are probable from standardisation of design, components and processes. Similarities are obvious with the manufacturing sector assembly lines. Process construction could also include client types where there is a relative balance between 'new build', maintenance, refurbishment or retrofit of existing assets. They may place large volumes of business into the industry, are ad hoc procurers of new build work, but deliver volume in the maintenance and refurbishment of existing assets.
- *Portfolio* construction, where clients have large, regular and ongoing investment programmes across a range of different project types. Portfolio construction will, however, involve a diverse range of needs in terms of technical requirements, degree of uniqueness or customisation as well as content. Regular spends will permit long term relationships with some suppliers. Clients involved in this type would be the MoD's Defence Estates organisation, BAA and Network Rail. This group could also include those client types where there is a significant volume

of 'new build' to the asset base as well as the maintenance, refurbishment or retrofit of existing assets.

A consolidated client typology is presented in Table 6.1. The public sector split into knowledgeable and less knowledgeable raises certain issues. For example, a small government science agency that has occupied its buildings, which are now out of date, for 30 years is likely to be an infrequent procurer.

In summary, clients to construction are heterogeneous. A client's value system comprises a number of interacting parts derived from its structure, cultural web, ownership characteristics and strategic management processes. Each client will have its own value system and drivers derived from its sector, organisational structure and functioning and the manner in which it approaches the industry, as identified above.

The discussion on the client value system will be extended further in the next section to differentiate between corporate and business value.

The client value system and portfolios of projects and single projects

A project or projects are an outcome of an organisation's strategic management process. They are the result of a series of business decisions made by senior managers within an organisation. Some projects may result in 'organisational projects', for example process simplification or re-engineering. Depending on the type of project and the client, others may result in projects requiring the creation of a physical asset such as roads, offices, hospitals or water treatment plants where profit, financial benefits to the country, or social need may be a driver, or a combination. The process is demonstrated in Fig. 6.2.

Programme management is as a term used increasingly to identify a set of management skills linking the strategic management process with the management of multiple projects in a consistent, coherent and integrated manner. An analogous term is a portfolio of projects. Project programme or project portfolio management is particularly pertinent in large corporate organisations where numerous projects can come on stream and need to be managed as a collective. Programme management is an integrating business function to support the core business of an organisation. As a business function it must have a demonstrable cost, quality and time benefit and has developed in response to the widespread use of projects as a means of realising strategic change. To be effective programme management requires the right projects to be managed, selected and coordinated and appropriate scarce resources allocated as and when necessary. It also requires appropriate monitoring and control procedures.

Depending on the organisational structure of the client, there will be a requirement to align projects with corporate and/or business unit missions and objectives to achieve value for money. Bell coined the term the 'value thread'[35], the idea that value must be transmitted, transformed and maintained either through a project network or a single project to ensure that value for money is

Table 6.1 Client impacts on the construction industry; adapted from Male (2002)[34].

Client type	Private sector						Public sector			
	Knowledgeable Regular procurers			Less knowledgeable Infrequent procurers			Knowledgeable Regular procurers		Less knowledgeable Infrequent procurers	
	Consumer clients: large owner occupiers	Consumer clients: small owner occupiers	Speculative developers	Consumer clients: large owner occupiers	Consumer clients: small owner occupiers	Speculative developers	Consumer clients: large owner occupiers	Consumer clients: small owner occupiers	Consumer clients: large owner occupiers	Consumer clients: small owner occupiers
Response to the industry										
Unique	—	—	—	✓	—	—	—	—	✓	✓
Customised	—	✓	✓	✓	✓	—	—	✓	NA	NA
Process	✓	✓	—	—	—	—	✓	✓	NA	NA
Portfolio	✓	—	NA	NA	—	—	✓	—	NA	NA

Note: The ✓ denotes that this is the *probable* occurrence. NA indicates *no* occurrence and a — indicates a *possible but unlikely* occurrence.

Value Context 161

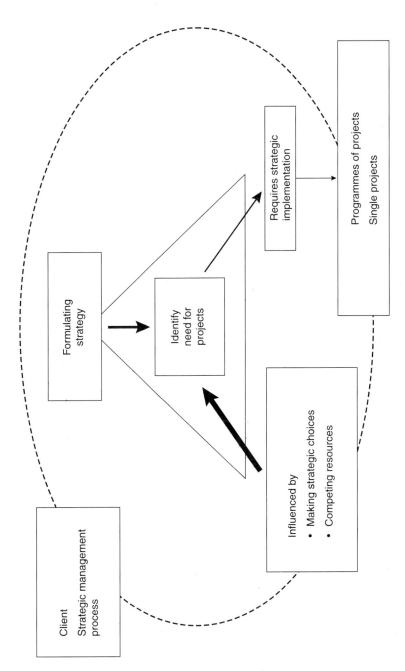

Fig. 6.2 The strategic management process and projects.

obtained as an output of the client organisation's strategic management process. In terms of client value systems stemming from the strategic management process[36], *corporate value* defines the requirements that exist for projects at corporate level within a client organisation with a diverse organisational structure. This may often reflect requirements that stem from a number of discrete business units or across a number of high profile corporate projects that are initiated and managed at the corporate level of the organisation. At this level the important requirement is to align projects with corporate and/or business unit missions and objectives. *Business value* defines the requirements that exist at business unit level for projects, or at the level of a single business entity that does not form part of a larger corporate organisation. Business units will be much more focused on industries, market sectors or particular products.

As an extension of corporate and business value it is also useful to think in terms of *programme value* and *project value*. The concept of the *project value chain* is introduced here to reflect the linkage between the organisation's strategic management process, projects as an outcome of that process and their need to be implemented as effectively and efficiently as possible. The project value chain encompasses both project programme delivery and individual project delivery. As a model it provides the conceptual value thread that links successive stages of project from inception, through development and into implementation. The project value chain is discussed in more detail in Chapter 7.

The next section brings the ideas discussed earlier together into a strategic value based model linking a client's strategic direction, physical assets, managing change and also project management.

6.4 A strategic value management model

As mentioned earlier, projects take place within the strategic context of the client's organisation and they are linked to a company's ongoing strategic direction. The forward direction of a client organisation will be worked out normally as part of a strategic plan for larger organisations made up of many businesses, or as a business plan for smaller, single entity businesses. It is often possible that due to the hierarchical structure of large corporate organisations a project or projects may start their life at the corporate strategic level and then be managed at individual business unit level. Equally, in larger organisations with more autonomous separate businesses, a project or projects may start life at this level with minimal if any input from the corporate strategic level. Much will depend on the size, complexity and strategic importance of the project(s) to the organisation, the level of investment required and the policy and operating procedures for handling such projects. This sets the parameters for the project development process.

It is also useful to think in terms of projects going through distinct phases. The 'strategic' phase operates at the interface between the strategic management process and projects or a project as an outcome of that process. The start and finish of the strategic phase is often difficult to identify[37]. There is a difference between the start of the strategic phase of a project and the start of the project. For example, with privately financed international water infrastructure projects the gestation of projects during the strategic phase can take many years. The start date occurs when the project becomes legal and has a customer identified for supply. The strategic phase will have started much earlier. Hillebrandt[38] reported a range of time periods for projects to emerge into the construction industry and Woodhead[39] indicated that time spans for the projects that he studied varied between six months and three years. Often projects will have gone through a variety of changes within the client organisation before they emerge as projects with a clear momentum ready for the industry to commence its work. The strategic phase of the project development process is therefore often 'messy', 'fuzzy' or 'ill defined' and can be difficult to identify down in terms of a clear start date. The 'tactical phase' is much more concerned with implementation. This will be elaborated on further in the next chapter, but for brevity both are encapsulated within the term 'strategic project management'.

Depending on the client's core 'business' activities, client organisations will experience change over time. At one end of a continuum change may be quite radical, as in rapid growth, downsizing or restructuring. At the other end of the continuum change may only deal with day-to-day activities within the business – operational change. Change may be externally driven, organisationally driven or technologically driven, or all of these. Consequently projects may have to adapt to technological and/or organisational change as part of their development process. This sets the context for the strategic and, subsequently, the tactical management process for project development and will become particularly important the longer the duration of the project.

Projects, depending on their type, may also contribute to the asset base of a company – such as, for example, a new or refurbished facility – or require the purchase of land or property. This also places projects in the context of asset management. Projects will have to be managed strategically and tactically. They deliver value to an organisation through the effective functioning at each of the project's successive stages – the project value chain, the subject of the next chapter. This will require the management of value in terms of the overall contribution that the project makes to the business. This is termed here the 'business project'. Business projects are also likely to require a technical solution, and value will need to be introduced into the technical solution. The contribution the technical delivery of the project will make to the business project is termed here the 'technical project'. Projects also have to be delivered, normally using external resources, and will involve an organisation contracting in third parties through the procurement process. Third parties will need to be linked contractually through the appropriate contract strategy. When in alignment these different parameters provide the overall 'fit' of projects strategically and determine their value contribution to the ongoing activities of the client organisation. Figure 6.3 highlights this schematically.

164 Frameworks of Value

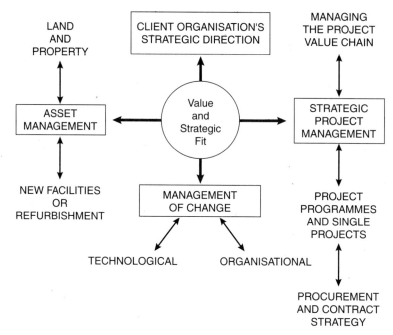

Fig. 6.3 The role of value, the strategic fit of projects and VM.

The next section will discuss, in the context of the foregoing, the role of value management and value engineering.

6.5 The role of value management and value engineering

Introduction: value drivers for projects

The previous sections have discussed the notion of value, indicating that it has both 'hard' and 'soft' attributes associated with it. The ideas of a 'value problem' and 'value systems' have been introduced. The term 'value thread' has been introduced and a number of concepts and models associated with that explored. A client typology has been introduced, resulting in an exploration of the client value system. Finally, value and the strategic fit of projects has been explored in the context of the client's system and a strategic value management model introduced.

Value management and value engineering

The exploration of value and its definition poses problems for value management as a discipline. It sits within a business and project context as set out in Fig. 6.3.

The authors contend that value management has a business focus and is strategic in nature whilst value engineering, a subset of value management, has a greater technical focus. In a construction industry context.

- Value management focuses on business project, which is the fundamental reason why a client organisation needs a project in the first place. The business project is expressed normally in a business case, justifying the investment for a project. The business project will be defined in terms of need, finance, returns, benefits, risks and time horizons. This is the strategic phase of a project.
- The technical project, the construction industry's response to that need, is the focus of value engineering studies. This corresponds to the tactical phase of the business project. The technical project will be defined much more in terms of technical specifications to meet that need.

A value management stance necessitates that the technical project should be in alignment with the business project to deliver value for money. Value management addresses the business project (which could include the contribution from a programme of projects) and the alignment of the technical project while value engineering is concerned with aligning correctly stages within the technical project to ensure the business project is delivered through an appropriate technical solution. Value management studies have to address not only the business and technical projects but also the objective and subjective elements of value as mentioned earlier.

Hence, the authors define value management as:

> The process by which the functional benefits of a project are made explicit and appraised consistent with a value system determined by the client.

Value engineering, on the other hand, is:

> The process of making explicit the functional benefits a client requires from the whole or parts of a project at an appropriate cost during design and construction.

Value engineering can also be:

> The process of identifying and eliminating unnecessary cost during design and construction.

These definitions highlight a number of important words and phrases. Value management is, first, a *process*. This implies that it has a series of steps. Second, it is about making choices on those components that comprise the value system of the client so that an appropriate decision can be made on the relative balance between, for example, capital expenditure and operational expenditure. This implies that it is not just about making choices on capital cost (the initial purchase cost) but about cost in use, which may include thinking about maintenance costs, operating costs and disposal costs. Making choices about cost could include:

- Deciding to reduce cost.
- Deciding to redistribute the way that cost is allocated.
- Deciding to increase cost.

Each is valid in terms of 'value' choices and establishes the decision framework on investment and what is affordable in an appropriate context. They can be seen as value strategies that need to be made explicit as part of a value study. It is important to emphasise in this context that neither value management nor value engineering is about cost cutting per se; this is only one of a potential number of value strategies that is available as part of a value study.

Third, the definitions highlight the importance of making *explicit* these choices. Value management is about bringing out clearly choices involving both the subjective and objective components to value. Fourth, the definition also mentions functional benefits. This draws out the economic, social and psychological dimension to value again and raises the question about who decides on the benefits and how.

Finally, value studies are team based because they involve resolving different and potentially competing and conflicting value systems coming together to address a particular value problem. They are about bringing the right people together at the right time either from within the client, from across the supply chain or both.

Furthermore a number of terms, such as 'value for money' (VfM) and 'best value', have taken root in the language of the construction industry. The latter, in particular, represents a new initiative by the UK government. If the definition of value set out here is correct, the initiative raises fundamental questions about who decides on what is value for money or what is best value and what is the appropriate frame of reference. Careful thought needs to be given to who is actually involved in making value choices, either for individuals, for groups, organisational units, organisations and firms, or as part of a political framework. It raises questions about politics; ethics; social norms and expectations; psychological processes; team dynamics; the pervasive influence of national and organisational cultures; the role of the State and political systems represented at national, regional and local levels; and, also the systems and procedures that are in place to inform decision makers about who to send and who not to send to value studies. It also raises questions about how long a value study should be, given that it involves as part of its process issues about values and value perspectives, resource allocation and what is an appropriate cost.

The book will explore these issues in much more detail later. Chapter 7 explores the concept of the project value chain.

6.6 References

1, 2. Bell, K. (1994) *The strategic management of projects to enhance value for money for BAA PLC*. PhD thesis, Heriot Watt University.
3. Gross, R. (1996) *Psychology: The Science of Mind and Behaviour*, 3rd edn. Hodder & Stoughton, Abingdon, Oxford.
4. Rosenberg, M. J. & Hovland, C. I. (1960) Cognitive, affective and behavioral components of attitude. In: *Attitude Organisation and Change: An Analysis of Consistency Among Attitude Components* (eds M. J. Rosenberg, C. I. Hovland, W. J. McGuire, R. P. Abelson & J. W. Brehm), Yale University Press, New Haven, Conn. Quoted in Gross, R. (1996) *Psychology: The Science of Mind and Behaviour*, 3rd edn, p. 434. Hodder & Stoughton, Abingdon, Oxford.
5. Gross, R. (1996) *Psychology: The Science of Mind and Behaviour*, 3rd edn. Hodder & Stoughton, Abingdon, Oxford.
6, 7. Yin, N. (1973) *The Study of Human Communication*. Bobbs-Merrill Company Inc, New York.
8. Allport, G., Vernon, P. & Lindsey, G. (1960). *A Study of Values*, 3rd edn. Houghton Miflin, Boston, Mass.
9. Gross, R. (1996) *Psychology: The Science of Mind and Behaviour*, 3rd edn. Hodder & Stoughton, Abingdon, Oxford.
10. Shillito, M. L. & De Marle, D. J. (1992) *Value: Its Measurement, Design and Management*. Wiley, Chichester.
11. Mudge, A. E. (1989) *Value Engineering: A Systematic Approach*. J. Pohl Associates, Pittsburgh.
12. Parker, D. E. (1977) *Value Engineering Theory*. Lawrence D. Miles Value Foundation, Washington D. C.
13. Dell'Isola, A., (1997) *Value Engineering: practical applications for design, construction, maintenance and operations*. R. S. Means Co, Kingston, Mass.
14. Mudge, A. E. (1989) *Value Engineering: A Systematic Approach*. J. Pohl Associates, Pittsburgh.
15. Fallon, C. (1980) *Value Analysis*, 2nd revised edn. Miles Value Foundation, Washington D. C.
16. Miles, L. D. (1989) *Techniques of Value Analysis and Engineering*, 3rd edn. Lawrence D. Miles Value Foundation, Washington D. C.
17. Scott, M. C. (1989) Value *Drivers: the Manager's Guide to Driving Corporate Value Creation*. Wiley, Chichester.
18. Porter, M. E. (1985) *Competitive Advantage: Creating and Sustaining Superior Performance*. The Free Press, New York.
19. Harvey, J. (1984) *Modern Economics*, 4th edn. Macmillan, Basingstoke.
20. BS EN 12973:2000. *Value Management*. British Standards Institution, London.
21. Checkland, P. B. (1981) Science and the systems movement. In: *Systems Behaviour* (Open Systems Group) 3rd edn, pp. 288–314. Harper & Row, London.

22. Male, S. P. (1991) Strategic management and competitive advantage in construction. In: *Competitive Advantage in Construction* (eds S. P. Male & R. K. Stocks), pp. 45–104. Butterworth-Heinemann, Oxford.
23, 24. Johnson, G. & Scholes, K. (1999) *Exploring Corporate Strategy: Text and Cases*, 5th edn. Prentice Hall Europe, Hemel Hempstead.
25. Langford, D. & Male, S. P. (2002) *Strategic Management in Construction*. Blackwell Publishing, Oxford.
26. Male, S. P. (1991) Strategic management and competitive advantage in construction. In: *Competitive Advantage in Construction* (eds S. P. Male & R. K. Stocks), pp. 45–104. Butterworth-Heinemann, Oxford.
27. Kay, J. (1993) *Foundations of Corporate Success: How Business Strategies Add Value*. Oxford University Press, Oxford.
28. Prahalad, C. R. & Hamel, G. (1991) The core competence of the corporation. In: *Strategy: Seeking and Securing Competitive Advantage* (eds C. A. Montgomery & M. E. Porter), pp. 277–300. Harvard Business Review, Boston, Mass.
29. Kay, J. (2000) Strategy and the Delusion of Grand Designs in Mastering Strategy, pp. 5–16. *Financial Times*.
30. Male, S. P. (2002) Supply chain management. In: *Engineering Project Management*, (ed N. J. Smith), 2nd edn. Blackwell Publishing Ltd, Oxford.
31, 32. *Croner's Management of Construction Projects*, July 1999, pp. 2-541-543 Croner, CCH Group Ltd, Kingston-upon-Thames.
33, 34. Male, S. P. (2002) Supply chain management. In: *Engineering Project Management*, (ed N. J. Smith), 2nd edn. Blackwell Publishing Ltd, Oxford.
35. Bell, K. (1994) The strategic management of projects to enhance value for money for BAA PLC. PhD thesis, Heriot Watt University.
36. Standing, N. (2000) *Value engineering and the contractor*. PhD thesis, University of Leeds.
37. Graham, M. (2001) *The strategic phase of privately financed infrastructure projects*. PhD thesis, University of Leeds.
38. Hillebrandt, P. M. (1984) *Economic Theory and the Construction Industry*, 2nd edn. Macmillan, Basingstoke.
39. Woodhead, R. (1999) The influence of paradigms and perspectives on the decision to build undertaken by large experienced clients of the UK construction industry. PhD thesis, University of Leeds.

7 The Project Value Chain

7.1 Introduction

This chapter builds on the previous one and draws together work that commenced on the strategic phase of projects and the project value chain[1]; the recognition of a two-stage briefing process[2]; and work that has been undertaken collaboratively with a number of researchers from industry[3,4,5] and academia[6,7]. The chapter also draws on an action research programme of studies that has been underway since the early 1990s. This has encompassed a series of value-for-money studies including delivering projects using value management and value engineering throughout the project life cycle; procurement studies that have encompassed PFI, prime contracting, partnering and other procurement routes; and finally organisational development projects that have involved setting up team structures, reviewing departmental organisational structures and introducing change into organisations. Finally, the chapter also integrates recent work undertaken by Male[8] on supply chain management.

The previous chapter introduced the concept of the project value chain. It is an extension of corporate and business value. It includes *programme value* (the management of programmes of projects) and, *project value* (the management of individual projects). Projects are an outcome of the strategic management process and comprise their own project value system. This stems from participants that are involved at each successive stage of the project. Just as a relay team has to pass on the baton from one person to another smoothly and efficiently to stand a chance of winning the race, the same principle operates within the project value system. Participants at each stage of the project life cycle have to pass on the 'baton' efficiently and effectively to those involved in the next stage. This creates the structure to the Project Value Chain. The project value chain encompasses both project programme delivery and individual project delivery and provides the conceptual value thread that links successive stages of project from inception, through development and into implementation.

This chapter presents the theoretical arguments behind value management as a project intervention methodology and process. The first section discusses VM in the context of projects.

7.2 Value managing projects

Managing a project necessarily relates to the management of information. In the later stages the information relates more to the function of space provision or more operational aspects. Value management discovers, structures and processes information held by a diverse number of stakeholders in a manner that permits key functions to be discovered, solutions found and actions planned. Using a number of structured techniques the value manager assists the project team to reach value-for-money solutions and plan actions for the efficient progress of the project.

The *Oxford English Dictionary* defines a project as being a plan, a scheme, or a course of action. Borjeson[9] defines a project as 'a temporary activity with defined goals and resources of its own, delimited from but highly dependent upon the regular activity'. Morris & Hough[10] define a project as 'an undertaking to achieve a specified objective, defined usually in terms of technical performance, budget and schedule'. Therefore a project is the 'investment of resource for return' where the investment is defined as being financial, manpower and/or material and the return commercial or social. This is a useful definition as it does not restrict the project to any particular sector.

The essence of all of these definitions is the recognition that the investment in a project is undertaken to add value to the core business of a client. The project has by definition a start date, a completion date, resources for its undertaking, a method for its smooth integration into the core business and, ideally, performance indicators which allow its impact on the core business to be measured. A significant danger of failure in respect of smooth integration and performance becomes apparent when the project value system is allowed to develop independently of the client value system. The relationship of the project to the core business is illustrated in Fig. 7.1. The client is only interested in the project from the perspective of increasing social or commercial wellbeing. The client's value system is also core business focused.

The method by which a client handles a project is generally related to the size

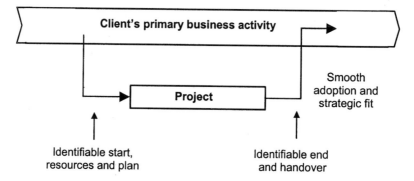

Fig. 7.1 The relationship of the project to the core business.

and complexity of the organisation. In the early stages of the development of a business, or in situations where the undertaking of a particular type of project is rare, the project represents such a large investment in relation to the turnover of the business or is so unusual that the project receives considerable management focus. In these situations, because executives have a hands on approach, the project value system is likely to be in tune with the client value system irrespective of whether both have been made explicit.

An organisation that undertakes projects as a matter of routine is likely to have put in place systems that administer projects. In these situations executives have a hands off approach and the project value system is more likely to be out of tune with the client value system stemming from structures and processes much deeper in the organisation. This danger is heightened when the client value system has not been made explicit and transferred to those responsible for administering the project. For example, a hotel owner with one hotel who is building a second hotel is likely to be highly involved in the project whereas the chief executive of a worldwide hotel chain is not likely to be involved in a project to procure an additional hotel. However, the core business of selling bedrooms is the same in each case.

Projects should not be confused with activities. Activities are the day-to-day processes necessary for the organisation to carry out its core business. In the hotel example above, each time a bedroom is cleaned is not a project but an activity necessary within the process of selling bedrooms.

The cyclical redecoration of bedrooms is an administered project whereas a project set up to examine and subsequently undertake a makeover of the whole hotel chain is a managed project. This difference is illustrated in Fig. 7.2.

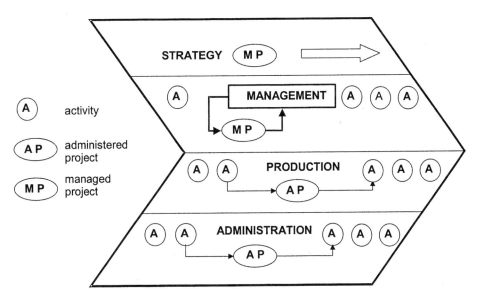

Fig. 7.2 Place of administered and managed projects within the organisation.

The generic stages in the development of projects

Four stages can be identified in a project's development.

Stage 1: strategic plan and business definition

As stated above, all projects are an outcome of the strategic management process and therefore can be traced back to that. The strategy may set up the requirement for an administered project, for example 'it is policy to refurbish our hotel bedrooms every three years', or a managed project may be prompted by such factors as a marketing opportunity, new legislation impacting the organisation's activities, a fall in the share price, a merger with another organisation, etc. The project at this stage exists as a single sentence mission statement. For example:

- A local authority, following a survey of parents, makes a policy decision to offer all parents a coordinated service for children within the age range 0 to 8 years.
- A car manufacturer sees an opportunity for the production of a new form of utility vehicle.
- An insurance company decides to consolidate all of its administration into a single head office building.

Stage 2: project planning and the establishment of systems

Once the project has been recognised and a decision taken to proceed a project planning stage is implemented. Activity for the example projects at stage 2 includes:

- The local authority begins to collect data and set up a project committee to investigate the components of such an early childhood service.
- The car manufacturer similarly sets up a study team to begin to formulate the performance requirements of the new vehicle strategy.
- The insurance company, either through a consultant or an in-house project manager, compiles a brief for the proposed building.

Stage 3: tactical design of the component parts of the project

At this stage all work necessary to bring the project into reality is undertaken and the majority of the capital investment is expended. In the example projects:

- The local authority will formalise the project team, involving all members of the appropriate public service sectors to tactically plan for the operation of the new service.
- The car manufacturer commences the design of the new vehicle, the machine tools and other facilities required in carrying out the manufacturing of the final product.

- The insurance company commissions the design and construction of the new building.

Stage 4: acceptance into core business

The final stage in the development of the project is the integration of the project into core business through service provision. In the example projects:

- The local authority accepts the output of the project team, which is disbanded. The new service goes into operation.
- The car manufacturer commences the production of the new vehicle.
- The insurance company occupies the new building.

An opportunity to add value to a project through value management occurs at each stage of its development but the effectiveness of value management is greater at the earlier stages of the project's development, as illustrated by the lever of value in Fig. 7.3. The concept is illustrated with reference to the example of the car manufacturer used above.

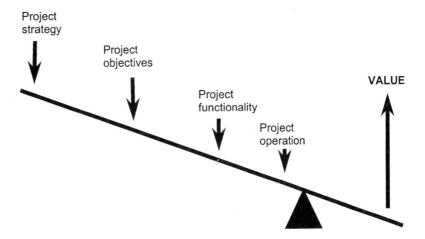

Fig. 7.3 The lever of value.

At the strategic stage the project team re-examines the basic requirement for a new utility vehicle, inviting to a workshop board of directors and senior management representatives. The client value system is discovered to place great emphasis on low capital investment, sensitivity to the environment and very high esteem. At the meeting the profitability of the project is judged marginal unless there is a high usage of existing components and/or the attractiveness of the vehicle is targeted at a wider customer profile than originally anticipated. The target for the use of existing components in the vehicle is set at 90% and the customer profile is referred back to marketing but with a directive to consider making the vehicle attractive to the 17 to 25 age group.

At project planning stage the project team consider the performance requirements for the vehicle, inviting to the workshop key supply chain members, senior salesmen from three main dealers and two insurance underwriters. A performance specification is derived for a vehicle that requires the minimum of re-tooling, whose characteristics place it in a low insurance group, with a high level of product differentiation through dealer fitted options. The resulting specification is audited against the client value system.

The tactical design of the component parts of the project is preceded by a workshop attended by the project team, the vehicle designers, all tier 1 suppliers of components and the production planning team. Concept designs for the vehicle were developed before the workshop to give an indication of basic shapes. The team derives a proposal based on an existing vehicle platform with an innovative space frame supporting a range of plastic body parts. A high level of prefabrication is to take place at tier 1 supplier level. A number of value engineering workshops take place at this stage to refine the production design and eliminate unnecessary cost. At the beginning and end of each workshop the client's value system is revisited and decisions made are checked.

At the assembly stage value engineering workshops involving operatives and production managers make adjustments to the vehicle production to remove unnecessary costs.

The lever of value clearly illustrates that effort applied at the strategic stage can have a dramatically higher impact on core business than equivalent effort applied at the operations stage. Further, the project's value system becomes more embedded as the process evolves and less able to be tuned to the client's value system. By the time the project is at the final design stage, irrespective of whether any implementation has taken place, it is virtually impossible to change the project's value system.

The remainder of this chapter develops ideas surrounding the project value chain in more detail.

7.3 The project value chain and the client value system

This section sets out the principle that the project value chain stems from activities deep within the client's strategic management process. It utilises ideas from business strategy and links these to the evolution of a project as an outcome of business strategy decisions.

The concept of the value chain stemmed from the thinking of Michael Porter[11] on business strategy and competitive advantage. He argued that a business organisation gains competitive advantage from the way that it structures, links and manages strategically important internal and external activities. Porter termed this the 'value chain'. Value chain activities provide the organisational infrastructure from which it creates value for the customers from its products or services.

In general, the concept of the value chain has been applied to business entities. However, it has been extended to projects Kelly & Male[12]. Building on the idea of the value chain from the business strategy field, proposed using the concept of the 'project as a value chain' within a value management framework. This permitted projects to be looked at as a series of linked, value adding activities and assisted in understanding a client's requirements at the strategic and tactical stages of a project. Bell[13] advanced the concept further within a project management context when she analysed project development and delivery within a large, volume procuring, owner-occupier client involved in managing major projects. The Reading Construction Forum document *Value for Money*[14] utilised the concept of value chains to prescribe the notion of adding value into the project through each stage of the chain. The document does not, however, describe how it is managed as a project value chain. Standing[15] extended the concept by looking at projects holistically which involved considering at an early stage such factors as the method of construction and procurement systems.

Standing[16] argues that the project value chain encompasses three distinct but interacting major value systems:

- *The client value system* which impacts the strategic phase – discussed in the next chapter.
- *The multi-value system* evident in the tactical phase – to be discussed below.
- *The user value system* evident in the operational phase – to be discussed below.

These three interacting value systems reflect major value transition points in any project. The project value chain is detailed in Fig. 7.4. It consists of first, the project programme level when appropriate, with a portfolio of projects to be managed, and second the individual project level. A sequence of value transitions can be delineated. These are:

- Corporate value.
- Business value.
- Project programme value.
- Project value, comprising:
    - Feasibility value.
    - Design value.
    - Construction value.
    - Commissioning value.
- Operational value.

In the case of an owner-occupier, client corporate/business value and operational value co-exist provided the project has been delivered effectively and efficiently.

The project value chain is therefore a multi-layered system of strategically linked activities potentially starting at programme level, for large corporate

176 Frameworks of Value

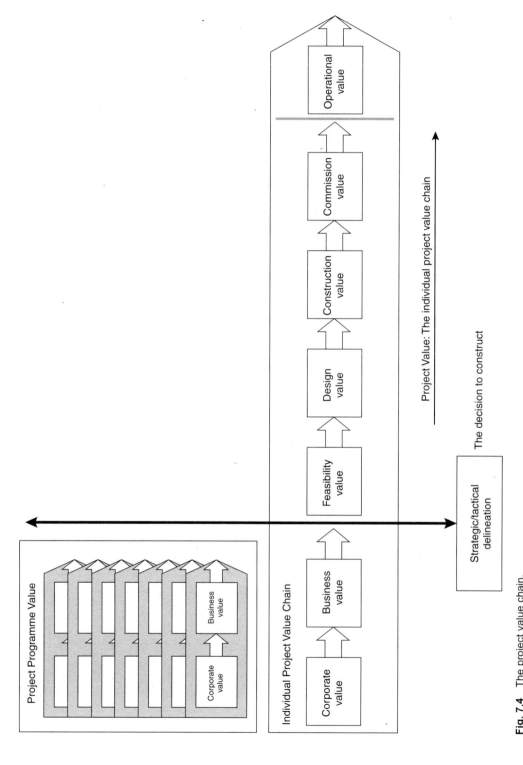

Fig. 7.4 The project value chain.
Source: Figure 2.4 in Standing (2001) *Value Management Incentive Programme*[17]. Reprinted with permission.

clients with numerous projects, and, subsequently moving to single project level. At programme level, the programme as a mechanism for implementing strategic change across the organisation must add value. Equally each project, process and phase within the programme must also add value for the delivery process to meet the strategic requirements of the client. Value management may address competition between projects for investment or how best to manage them as a holistic framework of projects. Equally the project value chain is also reflected strategically in linked value adding activities within a single project delivery throughout its life cycle.

From a project programme or portfolio perspective the strategic phase of a programme will encompass all competing networked projects. In such a situation single project delivery is therefore a tactical issue at programme or portfolio level. Equally single projects, delivered either within a portfolio or on their own, will have a strategic and tactical phase. Bell[18] proposed that a value hierarchy exists at programme/portfolio level since it will comprise projects at different stages of their life cycles. A hierarchy of value transitions for each project from concept through to use will also exist. This is demonstrated in Fig. 7.5.

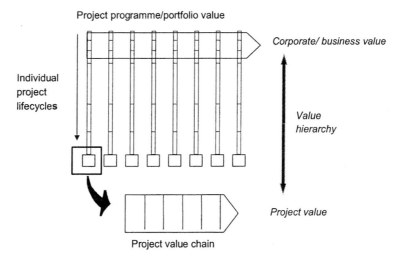

Fig. 7.5 Programme/portfolio level value.
Source: Bell (1994)[19].

7.4 The value thread within the single project value chain

This section introduces the idea that the transference of value through the different project stages creates a project value chain. There is the inherent potential within the project value chain for the transference of value to be successful or unsuccessful. The idea of a 'value thread' is introduced as an analogy to indicate the fragility of this transference within project activities.

The project value chain, forming part of an organisation's value chain where project activities are superimposed on the organisation's normal operations, leads to the concept that a project adds value to the organisation through its own processes. There are two primary transition points in the project value chain. The first is the decision to construct and the second is the handover of the completed physical asset into the operational domain. There are additional supporting transitional points as different organisations become involved. Discontinuities can occur as a result of changes in values due to the influence of individuals and teams from those organisations involved, consequently a different focus is being applied to the project.

The client value chain is concerned with a project, or projects, constructed to meet a business objective, or perhaps a social objective or a combination of both, depending on the type of client. The decision to construct is the point at which the client effectively outsources the business project to the construction industry in the form of a technical project to meet that need. The consequence is a requirement to ensure that the alignment of the different organisational value chains involved in the project process forms a holistic value driven, project focused chain working to the benefit of the client. Standing has viewed the project value chain as being subdivided into three distinct value systems:

- *The client value system* will comprise its own value chain for delivering the client's product or service. The client's value system creates demand for the construction industry, is business focused and is concerned with achieving corporate or business value from projects.
- *The multi-value system* comprises the construction industry supply chain and other external agencies such as planning departments, English Heritage, etc. The multi-value system is there to deliver the client's requirements as a technical response to the business project and also to ensure that various interested parties have an input into the process. However, different foci will be involved, for example:
  - ☐ The designer's value chain, as shown in Fig. 7.6, is design focused. It comprises numerous organisations involved in creating design value.
  - ☐ The contractor's value chain, as shown in Fig. 7.6, is task focused. It, however, comprises a myriad of suppliers and sub-contractors.
  - ☐ The operational value chain reverts to meeting the client's original business need through users. The hope is that the construction supply chain has worked effectively to deliver what the client, as customer, requires, i.e. fitness for purpose. Also, depending on the method of procurement adopted, the operational value chain may include third party operators contracted in to operate the physical asset on behalf of the client.
- *The user value system* comes into play throughout the operational life of the physical asset.

In drawing the foregoing together the next section explores the decision to construct/decision to build to provide a more generic framework for considering

The Project Value Chain 179

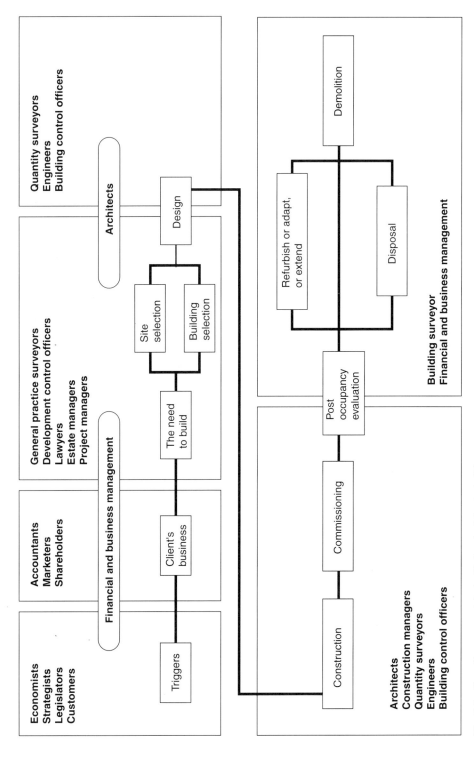

Fig. 7.6 Paradigms associated with the decision to build process.
Source: Adapted from Woodhead (1999).

value systems, value chains and the project value chain. It indicates the levels of complexity that can creep into the project value chain as a supply network.

7.5 The decision to construct/decision to build

As indicated above, the decision to build/construct is an important strategic value event in the project value chain at which point the project is effectively outsourced to the construction industry[20]. For example, a client can outsource the whole of the project value chain as with PFI or a client may have an in-house design capability and outsource construction. The decision to construct represents a business commitment that a project requiring some form of built solution is the right one and capital funding is being made available for further investigation and delivery. Corporate sanctioning is a key demarcation point. The impact of the multi-value system has been investigated both by Moussa[21] for project portfolios and by Woodhead[22] for single projects. The latter will be discussed first.

The value thread: the impact of paradigms and perspectives on the multi-value system

Woodhead[23] built on the work of Billelo[24]. The former undertook an in-depth analysis of the decision to build and uncovered an array of different influences that affect the process. He termed these influences paradigms and perspectives. Paradigms are the rules, expectations, values and codes of practice that are either explicitly or implicitly part of the professions that play a part within the decision to build process. Thus, a paradigm is common to a group and can be used to identify professions, representing their shared reality and their approaches to decision making. Perspectives refer to a particular view that can exist within any given paradigm. Woodhead argued that paradigms and perspectives are often reinforced by the professional chartered institutions, who codify standards and expectations or reinforce them through vocational study at universities. This propagates professional values and paradigms, as well as teaches a number of perspectives, or alternative views, within each paradigm. Figure 7.6 presents the array of paradigms that can influence the decision to build process

The following paradigms and perspectives have been identified in the decision to build process[26]:

- The capital investment paradigm.
- The cost benefit analysis paradigm.
- The financial paradigm.
- The strategic paradigm.
- The marketing paradigm.
- Organisational perspectives.
- Management perspectives.

- The property development paradigm.
- The planning permission paradigm.
- The preliminary design paradigm.

In addition Woodhead identified that paradigms and perspectives can influence the decision to build process in different ways. There are:

- Those that influence the process of the decision to build.
- Those that influence the content of the decision to build process.
- Those that intrude or are imposed on the decision to build process by external agents.

During the decision to build some paradigms and perspectives will be more dominant than others and conflict can stem from the influence of them competing. Thus they structure the process of the decision to build and dictate the content. Standing[27] provides a useful diagram that summarises the different value systems – or paradigms and perspectives – that can influence the project delivery process (Fig. 7.7).

The value thread and portfolios of projects: lean construction thinking

Moussa[28] investigated requirements across a range of European and North American airport clients. Lean manufacturing philosophy lies at the heart of her proposed framework for lean construction. She defined lean construction as:

> 'a holistic approach to project delivery focused on maximising value for money to the client through the use of innovative approaches to design, supply, and construction activities'.

Large, regularly procuring clients who are more likely to have an on-going programme of capital investment are in a better position to drive this process forward. They can provide the continuity of work and stability required to integrate the supply chain and practice lean thinking. Implementation of this philosophy requires a number of basic building blocks to form a coherent whole in order for it to succeed. These are set out in Fig. 7.8 and are typical of the requirements for collaborative working in projects now being advocated.

Lean thinking, when applied to construction, requires a number of underlying principles[30]:

- The link between the project programme/portfolio development cycle and the individual project development process permits/assists the formation of long term relationships between clients and the supply chain. A multi-disciplinary approach to projects can also occur much more easily since all project participants can be present at the very early stages.

182 Frameworks of Value

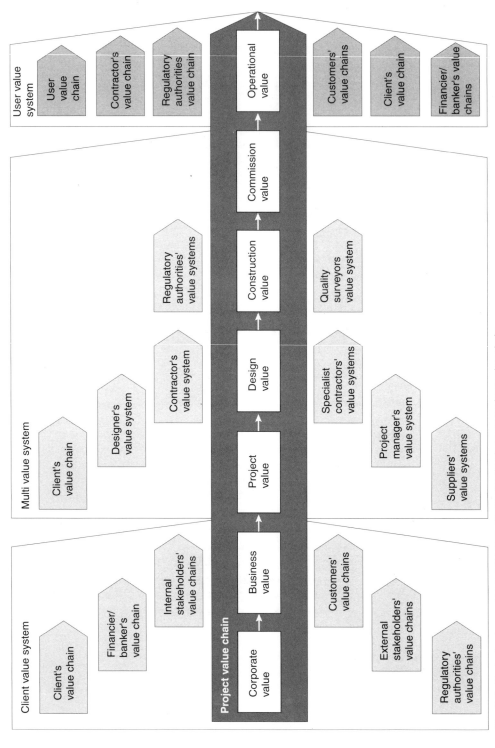

Fig. 7.7 Typical value systems and value chains that impinge on project value chain.
Source: Standing (2000)[27].

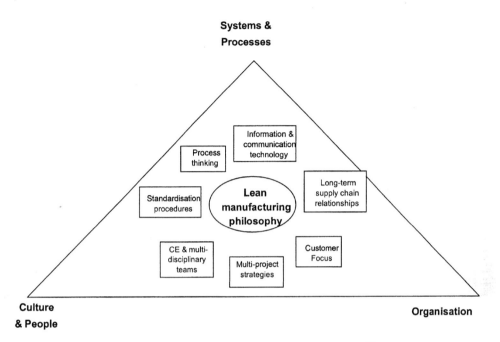

Fig. 7.8 Lean construction and portfolios of projects. *Source:* Moussa (2000)[29].

- The value hierarchy created from embedding single project processes within a higher multi-project portfolio process offers a strategic view of single projects. Through a process perspective this permits a focus on the repetitive aspects allowing continuous improvements to occur.
- Unifying separate design and construction stages within single project processes into one covering integrated design and production reflects a migration towards concurrent engineering and lean thinking. This permits moving towards prefabrication, modularisation and pre-assembly, minimising waste and reducing time spent on site.
- Benchmarking and performance measures initiated at the multi-project strategy stage, assessed and fed back to the process throughout the life cycle, and a specific emphasis on learning and the building of a client knowledge base, reflect a focus on process thinking and not simply a focus on single projects.
- Use of knowledge management in the last stage of the project and establishing dedicated documenting processes and product information promote waste reduction and the continuous improvement ethos of lean thinking.
- The incidence of a product development stage, where standard designs are commenced prior to the design processes at the single project level, permits standardisation of products and continuous improvement of design solutions. It also facilitates design for constructability and manufacturability and permits reduction of variability and waste.

- An emphasis on front-end planning by setting a project programme or single project strategy and performance targets, establishing relationships with the supply chain and product development.

The project development process addresses six stages at two different levels, namely the single and the multi-project levels. These stages reflect the changes required to achieve a lean approach to project delivery:

- *Stage 1* comprises creating a multi-project process, providing direction to the process, and defining and selecting a set of productivity and performance measurements.
- *Stage 2* comprises integrating and co-ordinating the supply chain involved in project development activities.
- *Stage 3* involves defining projects, investigating their viability and undertaking pre-planning activities.
- *Stage 4* comprises managing the design efforts so they converge to achieve business purposes as effectively and efficiently as possible.
- *Stage 5* comprises managing the delivery process to achieve the standards of performance required.
- *Stage 6* comprises creating and improving the capabilities needed to continuously improve the process over time.

These stages fit within the overall project process as defined in Fig. 7.9. For presentational purposes, the stages are shown in a level and sequential format due to the progression of information that must occur. Activities occur, however, at two different levels – the multi-project and the single project levels – but they should remain as integrated as possible.

7.6 Supply chain management in construction

Construction supply chains bring to the development and completion of the final product for the client a range of skills, experience and knowledge. When used appropriately and at the right time they contribute and add value to project delivery. This section argues that it is important to understand the principles of supply chain management in order to know when to incorporate these skills into a VM study or in designing a VM strategy that requires their use.

There is considerable ambiguity about terms and definitions and the scope of the field of supply chain management[32]. Terms often encountered include the firm's internal 'value chain' and external 'value system'[33], 'value stream'[34], the 'extended enterprise', 'lean enterprise', supply network and 'virtual organisation'. However, the essence of supply chain management is that it is a strategic function of the organisation. It integrates external and internal activities necessary to manage the sourcing, acquisition and logistics of resources that are essential for

The Project Value Chain 185

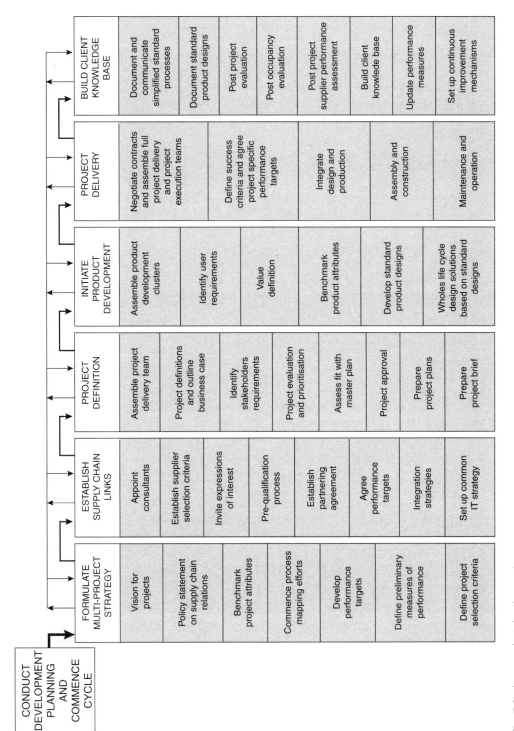

Fig. 7.9 Lean project development process.
Source: Moussa (2000)[31].

the organisation to produce products or services that add value to its customers. These terms collectively imply that the success of a organisation is not just confined to what is undertaken internally but is also dependant on the collective performance of firms within connected supply chains, often through the use of information technology as an enabler. There has been a shift in perspective from viewing the procurement of supplies as an operational activity in organisations into one that is now seen as strategic and linked to the long term survival of an organisation[35].

Currently supply chain management does not relate to a particular structure within an organisation but to a range of possible approaches that could be applied, such as an emphasis on progressive collaborative agreements or a portfolio of approaches, depending on the situation. 'Partnering' philosophies are typical. An increasingly important view is that different types of relationship may be equally appropriate in managing the supply chain. A more appropriate term is *demand chain management* since the chain should be driven by the market and not by suppliers[36]. Supply chain management, as the accepted term, is also a relatively new concept and is interrelated with sources of competitive advantage stemming from pitting supply chain against supply chain[37]. The client to construction is the motivating force behind demand, creating the markets and generating many of the pressures on the industry. They are the focus of the appropriate choice of the procurement system.

The two major influences on the supply chain are the client and the constructor. These are the two primary protagonists in contract with each other for the delivery of the product. Depending on the procurement route chosen, designers and other design team members may act as agents for the client as part of the demand chain. A number of procurement routes will, however, place them within the supply chain of the constructor. There are in effect two types of chain operating in the industry that make up the supply chain system for construction. One, generated by the client, is the *project-focused demand chain*; the other, managed by the constructor, has to handle multiple projects from a range of different clients – the constructor's *multi-project supply chain*.

Figure 7.10 indicates that, depending on the procurement route adopted, the client and design team generate the major cost commitment for most projects. However, collectively they are only responsible for approximately 15% of the client's expenditure on the project, primarily through design team fees. The constructor carries out the lion's share of expenditure on the project, approximately 85% of the project cost. Yet depending on the procurement route adopted, the constructor may be cut off from a direct influence on client and design team thinking, including influencing their commitment of cost in the early stages of projects. When looking at a project from a client and value-for-money perspective certain procurement routes preclude the constructor's knowledge, including their supply chain member's expertise from being accessed to the benefit of the project and potentially adding value much earlier in the project process. The next section looks at the impact of the procurement system on the project value chain.

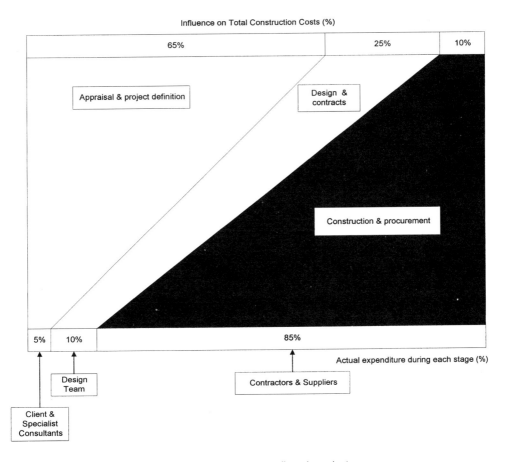

Fig. 7.10 Schematic comparing cost committed to expenditure in projects.
Source: Adapted from Fig. 3.4 in Standing (2001) *Value Management Incentive Programme*. Reprinted with permission.

Choice of procurement and the project value chain

The choice of procurement route is fundamental to setting up the parameters for delivering value through the project process. VM studies can often assist in deciding on procurement strategies or, depending on the need for a study, assist in sorting out the difficulties raised by the choice of procurement route. This section reviews the impact of procurement choice on the delivery of value.

Bell[39], Standing[40] and Moussa[41] contend that procurement systems need to take account of the portfolio and single project value chains from the client's perspective. As indicated above, procurement and contract strategies are important strategic decisions that can maintain the alignment of the project value chain or create barriers or discontinuities within it by facilitating or denying access to appropriate knowledge, skills and expertise regardless of where it is found within the supply chain. For example, at transitional points in the project value chain, such as the 'decision to build' and the 'operational user interface', discontinuities could occur due to changes in value systems present or a different focus being applied to the project. Equally, as a project(s) goes through each phase of its development from concept to use other value transition points occur, for example at feasibility stage, design stage (and even within the design stage), at construction stage, etc.

Providing the project value chain remains in alignment as a series of inputs and outputs, each should be creating value for the client[42], adding value until the complete project forms an asset for the client's organisation to meet a corporate need. However, complexity is added to the project value chain when other value systems impart skills and knowledge into it or create barriers to its effective operation as an integrated system working for and on behalf of the client.

Figure 7.11 provides a schematic overlay comparing some of the major procurement systems with the project value chain concept. There are a number of important issues that affect the integrity of the project value chain:

(1) Is the client able to transfer its own value system into documents and through dialogue with the supply chain in a manner that permits the multi-value system to understand this and translate it correctly? Examples include client output, specifications for PFI or prime contracts.
(2) Are members of the supply chain able to translate their own understanding of the client value system to other members, either operating concurrently or sequentially as part of the chain? Furthermore, some procurement routes detract from any meaningful contact being possible with the client.
(3) At important value transition points do interfaces work efficiently and effectively to transfer knowledge of the client's value system deeper into the chain?

The effectiveness of a procurement system is the extent to which is permits value transition points to work effectively. Schematically, those procurement systems at the top of the diagram provide potentially more opportunities to maintain the integrity of the project value chain since an increased number of discrete activities come under one umbrella organisation for single point delivery. There is one caveat, however: the project structure within the single point delivery organisation also has to work effectively. While the project remains within the client value system value should be maintained internally, although once transferred into the multi-value system through procurement there is a potential for loss in client value.

The contractor-led procurement systems, including PFI and prime contracting, where the contractor offers a 'one stop shop' service to a greater or lesser extent, have the potential for increased integration within the project value chain, depending on the method of tender for supply chain expertise. On the other hand, profession-led design procurement systems involve additional interfaces and provide more opportunities for disruption to the project value chain. Under

The Project Value Chain 189

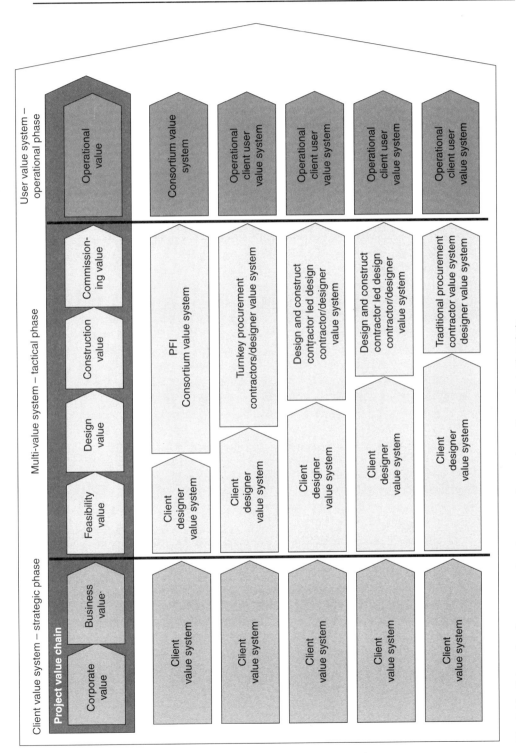

Fig. 7.11 Schematic of procurement systems superimposed over the project value chain.
Source: Adapted from Fig. 3.3 in Standing (2001) *Value Management Incentive Programme*[43]. Reprinted with permission.

profession-led systems any change by the contractor must have the client's approval since they are the only party that can sanction change. Mechanism permitting a change to the design by the contractor could occur but at a cost.

The management forms of procurement lie somewhere in the middle of the schematic. They permit increased involvement of construction knowledge earlier in the process. They are, however, essentially profession-led routes with a consequent increase in the number of interfaces.

The use of the PFI procurement system should, in theory, permit minimal disturbance to the project value chain especially if the client has defined correctly the output specification. Unlike other procurement routes, PFI has a 'double edge' since the consortium have to operate what they have designed and built, potentially for anything up to 30 years. Their focus should be on ensuring continuity and integrity of delivery throughout the process. The PFI procurement system, prior to the introduction of prime contracting, is the only system where the value thread could be maintained into the user value system providing one of the greatest opportunities for enhancing the impact of demand on supply chain management. Consortia have to balance the anticipated demands of the user with the correct ratio of operational and capital expenditure to maximise the return on the investment. Research evidence suggests that this is not being realised by the industry to date.

Turnkey procurement has similarities with PFI but, unlike the latter, does not have the additional requirement and liability for operating the physical asset. The contractual positioning and role of the designers will alter the impact on the project value chain. If the contractor employs the designers in house there should be increased alignment in the project value chain provided the project structure accommodates this. However, under turnkey, where the designer is independent of the contractor, an additional value system is imposed.

The traditional procurement route, at the bottom of the schematic, is probably the most disruptive to the project value chain since single stage competitive tendering, if adopted, occurs at the transition point between design value and construction value. Two-stage tendering, overlaid onto the traditional procurement route does, however, have the capability of bringing construction expertise into the project much earlier. Therefore there is also an interaction between procurement method and choice of tendering strategy in terms of impact on the project value chain. The value chain and value thread are potentially more robust when partnering structures are overlaid on the above. They permit different paradigms and perspectives to interact more effectively.

7.7 Creating value opportunities in the project value chain: value management (VM) and value engineering (VE)

This section explores how the VM process brings different value systems together with a view to moving a project forward.

The Project Value Chain

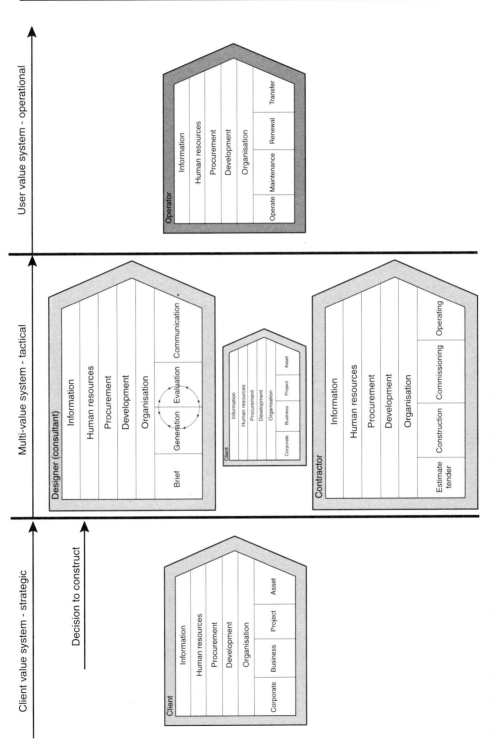

Fig. 7.12 Alignments in the project value chain (adapted from Porter, M.).
Source: Fig. 8.1, Standing (2000)[44].

The previous chapter introduced value management as a team based activity concerned with making explicit the package of whole life benefits a client is seeking from a project or projects at the appropriate cost. Value engineering, as a subset of value management, is a team based activity concerned with making explicit the package of whole life benefits that a client is seeking from the technical delivery of the project during design and construction. Value management is about articulating clearly the business benefits sought from a project. Value engineering ensures that these are delivered in design and construction. Value management and value engineering are project methodologies for aligning or re-aligning the project value chain. To achieve this, teams across the project value chain need to be brought together to ensure connectivity within the value thread. Figure 7.12 illustrates the point.

The choice of procurement route is a strategic value decision. It can, as indicated above, align or disaggregate the project value chain to a greater or lesser extent. Value management and value engineering, by bringing the right people together at the right time regardless of where they are located in the supply chain, is an integrating function within the project value chain. Kelly & Male in their first study into the subject conducted for the RICS noted that VM/VE has the potential to integrate project participants and process more effectively than any other mechanism allied to procurement. VM and VE, depending on whether they are used proactively or reactively as part of project delivery, can ally or re-align the project value chain through a series of value opportunity interventions. Given that the constructor's supply chain spends approximately 85% of the client's expenditure on projects, certain procurement routes prevent constructor supply chain input much earlier into the process. The effective use of value management as a method to integrate more effectively the project value chain is impacted directly by the procurement route chosen, or may require value management incentive programmes to assist the process where procurement routes prevent full alignment or re-alignment.

7.8 References

1. Male, S. P. & Kelly, J. R. (1992) Value management as a strategic management tool. In: *Value and the Client*, pp. 33–44. Royal Institution of Chartered Surveyors, London.
2. Kelly, J., MacPherson, S. & Male, S. P. (1992) *The Briefing Process: A Review and Critique*. Royal Institution of Chartered Surveyors, London. Kelly, J., Male, S. & MacPherson, S. (1993) *Value Management: A Proposed Practice Manual for the Briefing Process*. Royal Institution of Chartered Surveyors, London.
3. Bell, K. (1994) *The strategic management of projects to enhance value for money for BAA PLC*. PhD thesis, Heriot Watt University.

4. Standing, N. (2000) *Value engineering and the contractor*. PhD thesis, University of Leeds.
5. Graham, M. (2001) *The strategic phase of privately financed infrastructure projects*. PhD thesis, University of Leeds.
6. Woodhead, R. (1999) *The influence of paradigms and perspectives on the decision to build undertaken by large experienced clients of the UK construction industry*. PhD thesis, University of Leeds.
7. Moussa, N. (2000) *The application of lean manufacturing concepts to construction: a case study of airports as large, regular-procuring, private clients*. PhD thesis. University of Leeds.
8. Male, S. P. (2002) Supply chain management. In: *Engineering Project Management* (ed N. Smith), 3rd edn. Blackwell Science, Oxford.
9. Borjeson, L. (1976) *Management of Project Work*. The Swedish Agency for Administrative Development, Satskontoret, Gotab, Stockholm, Sweden.
10. Morris, P. J. W. & Hough. G. H. (1987) *The Anatomy of Major Projects: A Study of the Reality of Project Management*. Wiley, Chichester.
11. Porter, M. E. (1985) *Competitive Advantage*. Free Press, New York.
12. Male, S. P. & Kelly, J. R. (1992) Value management as a strategic management tool. In: *Value and the Client*, pp. 33–44. Royal Institution of Chartered Surveyors, London.
13. Bell, K. (1994) *The strategic management of projects to enhance value for money for BAA PLC*. PhD thesis, Heriot Watt University. Developed the concept further.
14. Reading Construction Forum (1996) *Value For Money*. Centre for Strategic Studies, Reading.
15. Male, S. P. & Kelly, J. R. (1992) Value management as a strategic management tool. In: *Value and the Client*, pp. 33–44. Royal Institution of Chartered Surveyors, London.
16. Standing, N. (2000) *Value engineering and the contractor*. PhD thesis, University of Leeds.
17. Standing, N. (2001) *Value Management Incentive Programme: innovations in delivering value*. Thomas Telford, London.
18, 19. Bell, K. (1994) *The strategic management of projects to enhance value for money for BAA PLC*. PhD thesis, Heriot Watt University.
20. Mitrovic, D. (1999) Winning alliances for large scale construction project on world market: profitable partnering in construction procurement. In: *Joint Symposium CIB W92, TG23*, pp. 258–63. E. and F. N. Spon, London.
21. Moussa, N. (2000) *The application of lean manufacturing concepts to construction: a case study of airports as large, regular-procuring, private clients*. PhD thesis. University of Leeds.
22, 23. Woodhead, R. (1999) *The influence of paradigms and perspectives on the decision to build undertaken by large experienced clients of the UK construction industry*. PhD thesis, University of Leeds.

24. Bilello, J. (1993) *Deciding to build: university organization and design of academic buildings*. PhD thesis, University of Maryland.
25, 26. Woodhead, R. (1999) *The influence of paradigms and perspectives on the decision to build undertaken by large experienced clients of the UK construction industry*. PhD thesis, University of Leeds.
27. Standing, N. (2000) *Value engineering and the contractor*. PhD thesis, University of Leeds.
28, 29, 30, 31. Moussa, N. (2000) *The application of lean manufacturing concepts to construction: a case study of airports as large, regular-procuring, private clients*. PhD thesis, University of Leeds.
32. Male, S. P. (2002) Supply chain management. In: *Engineering Project Management* (ed N. Smith), 3rd edn. Blackwell Publishing, Oxford.
33. Porter, M. E. (1985) *Competitive Advantage: Creating and Sustaining Superior Performance*. The Free Press, New York.
34. Hines, P., Lamming, R., Jones, D., Cousins, P. & Rich, N. (2000) *Value Stream Management: Strategy and Excellence in the Supply Chain*. Financial Times Prentice Hall. London.
35. Saunders, M. (1997) *Strategic Purchasing and Supply Chain Management*, 2nd edn. p. vii. *Financial Times* & Pitman Publishing, London.
36, 37. Christopher, M. (1998) *Logistics and Supply Chain Management: Strategies for Reducing Cost and Improving Service*, 2nd edn. pp 10 *Financial Times* & Pitman Publishing, London.
38. Standing, N. (2001) *Value Management Incentive Programme: innovations in delivering value*. Thomas Telford, London.
39. Bell, K. (1994) *The strategic management of projects to enhance value for money for BAA PLC*. PhD thesis, Heriot Watt University.
40. Standing, N. (2000) *Value engineering and the contractor*. PhD thesis, University of Leeds.
41. Moussa, N. (2000) *The application of lean manufacturing concepts to construction: a case study of airports as large, regular-procuring, private clients*. PhD thesis, University of Leeds.
42. Bell, K. (1994) *The strategic management of projects to enhance value for money for BAA PLC*. PhD thesis, Heriot Watt University.
43. Standing, N. (2001) *Value Management Incentive Programme: innovations in delivering value*. Thomas Telford, London.
44. Standing, N. (2000) *Value engineering and the contractor*. PhD thesis, University of Leeds.

8 Client Value Systems

8.1 Introduction

Value is often defined as a relationship between cost and quality. Burt[1] states that maximum value is achieved when the required level of quality is obtained at the least cost, the highest level of quality is achieved for a given cost, or from an optimum compromise between the two. This chapter explores the place of structured value criteria within an existing quality environment. It discusses the concept of quality and demonstrates that value management and value engineering are powerful methodologies when total quality management and quality assurance have exposed a project to change and/or to improve procedure. The chapter concludes by bringing together information from chapters in Part 2 to define the client value system using a structured technique.

8.2 Defining quality as part of value

Definitions of quality, even in textbooks and articles on the subject, are few in number. Bicheno[2] quotes Deming who states that quality can only be defined in terms of customer satisfaction. There is no absolute measure: two customers may perceive a product or service differently. Juran[3] defines quality as 'the totality of features and characteristics of a product or service that bear on its ability to satisfy stated needs or implied needs. Quality consists of freedom from defects.' This definition of quality is repeated in standards. Vorley[4] quotes Juran's definition of quality and comments that quality needs to cover more than function and that even if aesthetics is included in function there are many other factors to be considered, for example the method of distribution, initial and running costs, user awareness or knowledge.

In his extraordinary book *Zen and the Art of Motorcycle Maintenance* Pirsig[5] states:

'why does everybody see Quality differently?... Quality is shapeless, formless, indescribable. To see shapes and forms is to intellectualise. Quality is independent of any such shapes and forms. The names, the shapes and forms we give Quality depend only partly on the Quality. They also depend partly on the *a priori* images we have accumulated in our memory. We constantly seek to find, in the Quality event, analogies to our previous experiences... The reason people see quality differently is because they come to it with different sets of analogies.'

If this is true then it is logical to assume that individuals setting targets for attributes of quality must have sufficient experience or knowledge of an analogous event and have the tools and measures to assess that event. For example, someone working with polyester resins might be in a good position to be able to define adequate ventilation in a workshop where polyester resins were being used. However, without appropriate tools and measures they may be unable to specify a quality ventilation system.

To take this one step further, consider two hotels. The first hotel is sited on a busy arterial road leading from a major city. There is no air conditioning and many rooms have a tendency to overheat. Rooms are basically furnished and each has an en suite bathroom in reasonable condition but with some mould growth and evidence that the room has not been too well cleaned. The dining room and bar areas are reasonably comfortable and the quality of the food is good. The second hotel is also sited on a busy road but is fully air conditioned and the rooms are soundproof. Rooms are large and well furnished with modern furniture in good condition. The en suite bathroom is finished in marble, has a shower in addition to a bath and is immaculately clean. There is a choice of dining room and bar areas which are all interesting and comfortable and the food is very good.

The question arising from the logic of Pirsig is whether the specification writer who had only ever stayed in a hotel of the first type could envisage the second type. The answer is probably not unless other knowledge and experiences were brought to bear, e.g. illustrations or descriptions in books, television programmes, standards of cleanliness in other situations or experience of air conditioned offices in similar locations. This dilemma is illustrated in Fig. 8.1. On a continuum from the highest degree of excellence achievable to the lowest provision of the basic function every individual will create a frame of reference based upon experience and knowledge, represented by the circle on the scale. The individual's target level of quality will be within the frame of reference. There is a high probability that because every individual's knowledge and experience is different the frames of reference for all individuals will not be in the same place on the continuum.

The next logical question is what happens in the situation where the product/service receiver has more experience than the product/service specifier? The logical answer to this question is that there is a high probability of a poor quality product or poor quality service. A way around this dilemma is to ensure that the appropriate experience is brought to the writing of the specification and suitable models are available for the interrogation of the experience.

Bicheno[6] describes the Kano model developed by the Japanese quality guru

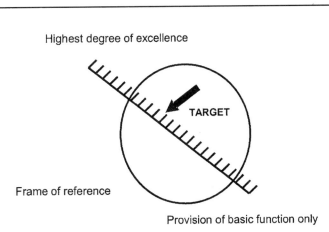

Fig. 8.1 The quality continuum.

Dr Noriaki Kano who states that maximum quality is attained when targeted characteristics are achieved and the customer is delighted. There are three variables within the model. These are 'basic factors', 'performance factors' and 'delighters', and all have a relationship to the presence of quality characteristics and customer satisfaction. These variables are included in the Kano model, illustrated in Fig. 8.2, and the quality matrix, Fig. 8.3.

In the Kano model a basic characteristic is expected to be present; the cus-

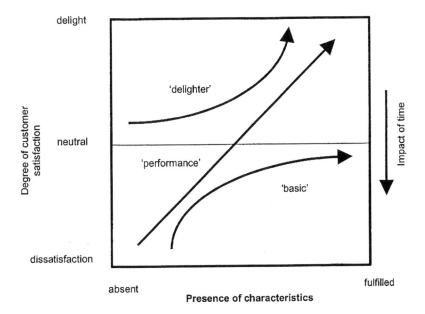

Fig. 8.2 The Kano model.
Source: adapted from Bicheno (2000) 75[7].

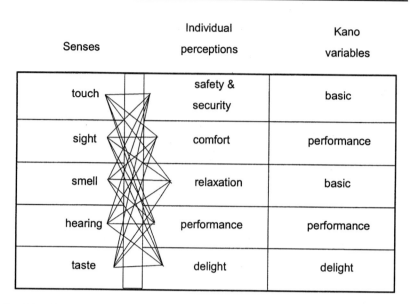

Fig. 8.3 The quality matrix.

tomer will be dissatisfied if it is absent and only neutral if the characteristic is completely fulfilled. The performance characteristic relates to the essential function. The customer will be more satisfied if higher levels of performance are achieved. The delighter is the unexpected extra characteristic that was not expected by the customer. There is, however, a time dimension to the model such that the three variables will tend to sink over time, i.e. what once delighted is now expected and higher levels of performance are always sought. For example, power steering on small cars as a standard feature once delighted customers but now power steering is expected as a basic characteristic and its absence would lead to dissatisfaction.

A suitable model for interrogation must address the means by which we determine quality, which is through the senses as illustrated in Fig. 8.3. The five senses of touch, sight, smell, hearing and taste lead to individual perceptions which are translated into 'states of personal awareness' of safety and security, comfort, relaxation, performance/excitement and delight. It is these that enable the individual to judge quality. In the synthesis of the quality continuum, the quality matrix and the Kano model it is deduced that the perception of quality is personal and is related to the stimulation of the senses.

As variables on the Kano model feelings can be determined as being basic, performance or delight. For example, having arrived in a hotel bedroom an individual will use the senses to determine if they are safe and secure. If they feel threatened they will be dissatisfied, hence safety and security is a basic function. It should be noted that safety and security is not a performance function since 100% safety and security is expected and thus its presence will be treated as neutral. Performance factors are based on a combination of knowledge and/or experience and can lead to either satisfaction or dissatisfaction. The delighters are the unexpected and will enhance customer satisfaction and potentially form the basis of a different set of value criteria for the future. Quality requires the provision of all basic functions at the required level and all performance functions at the highest level. Delight functions provide the unexpected value added.

The quality matrix and the Kano model present useful models to assist the quality specifier to understand the knowledge and experience of the customer.

Redefining quality

Summarising the above, 'quality' is defined here as the degree to which stated objectives, characteristics and/or attributes have been met. This is often associated with a degree of excellence and is the provision of all basic functions at the required level and all performance functions at the highest level. Delight functions provide the added value. Quality is thus is a construct of comparability.

The next section explores how quality is translated into systems that make quality explicit in tangible terms.

8.3 Quality systems

Two quality systems are explored: total quality management and quality assurance. Total quality management is described as an attitude that pervades the management culture of an organisation to provide the highest level of excellence in products and/or services. Quality assurance, on the other hand, is the management of defined, consistent, standards of products and/or services[8]. Quality assurance necessarily requires an understanding of the performance capability in the production of products or delivery of services, the setting of standards, the measurement of performance and the means of remedial action. The two systems will be discussed in turn.

Total quality management

Vorley[9] defines total quality management as the synthesis of the organisational, technical and cultural elements of a company. This attitude is made explicit in W. Edwards Deming's 14-points total quality management[10] listed:

(1) Create constancy of purpose for improvement of product and service. Deming's view is that management should focus their attention on quality rather than profit as profits follow as a natural consequence of an organisation targeting quality.
(2) Adopt the new philosophy. Deming's view is that mistakes raise costs and

cause delays whereas reliability in management and production reduces both costs and delays. It is not sufficient that defects are minimised but rather they should be eliminated.

(3) Cease dependence on mass inspection. Deming recognises that errors discovered by inspection imply that efficiency and effectiveness have already been compromised. The principle of quality by design is more efficient and enhances employee satisfaction through a job well done. An effect of quality by design is that the employee has to work against the design to produce poor quality.

(4) End the practice of awarding business on the price tag alone. This is a clear principle echoed by the Latham and Egan reports. Deming was an early advocate of measuring quality alongside cost in tendering.

(5) Constantly and forever improve the system of production and service by adopting the principle of continuous improvement.

(6) Institute modern methods of training on the job. This principle is echoed in current knowledge management thinking and relates to the notion that operatives and managers should constantly be updated with the knowledge and skills required to carry out their operations efficiently and effectively, and that where knowledge lies should be comprehensively mapped.

(7) Institute leadership. According to Deming leaders begin with the assumption that workers aim to do the best job they can and therefore the role of management is to help workers reach their full potential through training and empowerment. This culture is more effective than instruction by supervisors within a context of discipline, rewards and penalties.

(8) Drive out fear. Total quality management requires communication at all levels within the organisation such that no one within the organisation should feel constrained in pointing out non-value-adding activities.

(9) Break down barriers between staff areas. Deming considers barriers between functional departments to be counterproductive. Often recognised as 'silo management' and 'over the wall thinking', this attitude focuses attention on groups within silos carrying out their own activity and throwing their product over the wall to the next group, also contained within their isolated silo. More often recognised through the supply chain than within the organisation, this activity leads to poor value through a lack of focus on the final customer.

(10) Eliminate slogans, exhortations and targets for the workforce. Deming criticises companies that attempt to motivate employees through slogans and exhortations as in his view this frustrates employees.

(11) Eliminate numerical quotas. Deming advocates the removal of quotas because they encourage a focus on quantity at the expense of quality.

(12) Remove barriers to pride of workmanship. In Deming's view merit systems should be eliminated. If people inherently want to perform well they do not need incentive systems; rather they need to be facilitated in overcoming obstacles imposed by inadequacies in materials, equipment

and training. This raises fundamental issues for the basis on which collaborative procurement methods in construction are currently structured.

(13) Institute a vigorous programme of education and training. This must be seen as the understanding of the overall total quality management aspirations of the organisation as well as the training undertaken by managers and operatives to fulfil their work activity.

(14) Take action to accomplish the transformation. Deming's view is that if total quality management is to be instituted within an organisation as a culture then the whole organisation must work together to enable that to succeed.

The implementation of Total Quality Management as envisaged above relies upon making explicit the organisation's beliefs and culture or the organisation's value system. The next section reviews external measures or benchmarks of total quality.

Total quality award schemes

Total quality award schemes are worthy of examination from the perspective of the identity of the criteria and the weight given to each criterion when judging a quality organisation. The argument made later is that these criteria should be in harmony with the organisation's value system.

The Deming Prize

The earliest quality award scheme is the Deming prize first awarded at the first quality control annual conference in Japan in 1951. The conference and the awarding of the Deming prize is administered by the Union of Japanese Scientists and Engineers (JUSE).

The Baldridge National Quality Award

The Balridge National Quality Award was instituted in the USA in 1987 and named after Malcolm C. Baldridge in honour of President Reagan's Secretary of Commerce who died in a horse-riding accident. Companies apply to be assessed under seven categories:

(1) Leadership (100 points) assesses senior executive leadership, the quality values of the company, the management of quality and public responsibility.
(2) Information and analysis (70 points) examines the scope, validity, use and management of data and information, its adequacy and support for quality assessment.
(3) Strategic quality planning (60 points) is for achieving or retaining quality leadership in company planning.

(4) Human resource utilisation (150 points) examines the use of the full potential of the workforce including management.
(5) Quality assurance of products and services (140 points) examines statistical and procedural approaches to design and the production of goods and services.
(6) Quality results (180 points) examines the levels of quality improvement based upon objective measures derived from the analysis of customer requirements.
(7) Customer satisfaction (300 points) examines the company's knowledge of its customers and its responsiveness to those customers.

The European Quality Award (EFQM)

The European Quality Award was born out of recognition that public and commercial organisations have a responsibility in terms of the needs and expectations of their customers and stakeholders, workforce and local community. In 1988 the European Foundation for Quality Management was formed by 14 leading European companies. In 1992 the first European Quality Award was presented. The scheme is administered through national organisations that organise training and promote and manage the award. Over recent years, however, and particularly in local government, EFQM has become synonymous with self-assessment and unlike the ISO 9000 series is not based on certification.

The EFQM model consists of nine criteria and 32 sub-criteria that are split into two categories: 'enablers' and 'results.' The enablers can be viewed as the systems of what an organisation does, how it does it and future direction, looking at leadership and people processes, policy and strategy, and partnerships and resources. The results are what the organisation has accomplished in terms of achieving its key performance goals with respect to people, customers and society. A diagrammatic representation of the EFQM model is shown in Fig. 8.4. The percentages shown are those used for assessing applications for the European Quality Award; for self-assessment purposes organisations may chose their own weightings. The self-assessment is normally conducted using a weighted matrix of statements each of which attracts between 1 and 10 points. Those undertaking self-assessment align with the statement which they believe to be true in the context of their own organisation. It is generally presumed that the statements are inclusive of lower marked statements, i.e. to score 10 both the statement and the statements included below it are seen to be applicable. An example is shown in the Table 8.1.

The total score possible within the scheme is 1000 points. Companies scoring over 500 points are seen to be best in class.

The EFQM model (Fig. 8.4) forms a checklist of good practice that brings together various quality initiatives within a framework. It allows an organisation to measure itself on a regular basis by gathering evidence relating to the nine criteria prior to an assessment and identifying any improvement opportunities.

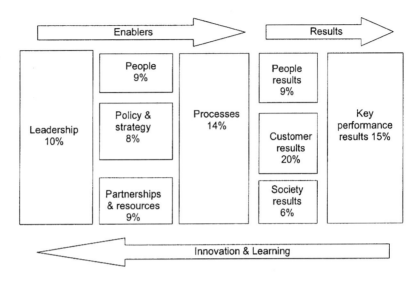

Fig. 8.4 The EFQM model (EFQM® and the model are registered trademarks of the European Foundation for Quality Management).

Table 8.1 EFQM self assessment statements.

Assessment statement: people	Points	Weight	Total score
Management at all levels uses an effective team approach combined with individual empowerment	10	9	90
Structured knowledge management with training is employed at all levels	5	9	45
Employees' opinions are sought and training given as necessary	1	9	9

Because TQM has a focus on organisation culture and value, there is a close relationship to value systems and value cultures.

Quality assurance

Juran[11] defines quality assurance as all those planned and systematic actions necessary to provide adequate confidence that a product or service will satisfy given requirements for quality. Quality assurance is the management of defined, consistent standards of products and/or services. QA systems can operate within a quality driven culture as TQM, but because it is focused on systems and procedures it can also act independently of such a culture.

Other quality schemes

Six Sigma

Six Sigma is a management tool derived from the statistical analysis of defects. The term 'six sigma' is a statistical goal for organisations, which means that products, processes or services will experience only 3.4 defects per million opportunities, or 99.99966% good. When an organisation achieves Six Sigma it is considered to be best in class. Motorola first used the term in its winning application for the 1998 Baldridge award although the practice of the methodology preceded this by a decade. Breyfogle[12] indicates that most organisations are thought to operate at about four sigma, which means that they have 1826 more defects per million opportunities and make 10% less in profit than those operating at six sigma.

Six Sigma is a measure of quality where a defect is defined as anything outside customer specifications. Based upon the premise that customers value a consistent, predictable service the main objective of Six Sigma is customer satisfaction through continuous improvement in quality with the goal of near perfection. If the organisation can measure how many defects are present in a product, process or service they can work out how to eradicate them and get as close to 'zero defects' as possible. The Six Sigma three elements of quality are the customer, the process and the employee.

Balanced scorecard

Based upon a seminal paper by Kaplan & Norton[13] the balanced scorecard is a strategic management approach described as both a management and a measurement system that allows an organisation to analyse their vision and strategy and consider the actions to be taken. The balanced scorecard views the organisation from four perspectives: innovation and learning, internal business perspective, the customer perspective and the financial perspective. Specific data is collected relating to the four perspectives with the aim of determining actual against planned performance.

Summary

To summarise, this section has reviewed and defined quality and reviewed different quality systems. Quality is an inherent component of value but it can easily be ill-defined, may not be made explicit and may not pervade the organisation. VM is culture focused and is therefore in close alignment with TQM, whereas QA is much more systems and procedures focused and stands alone. In this sense performance measures inherent within an organisation committed to TQM are more easily transposable into value criteria for use in VM.

By abstraction from the quality criteria, the raw data for a value system could incorporate:

- Culture in terms of leadership, human resource organisation and policy. Policy in this context would include environmental and other issues.
- Organisational processes incorporating flexibility, agility, stability, etc.
- A focus on the customer.
- Organisational attitudes to society and the neighbouring community.

8.4 Performance indicators

An important feature of all of the quality systems described above is the necessity to be able to establish meaningful criteria against which performance can be measured. The measurement criteria for total quality management are different from quality assurance.

The construction industry has spawned a number of such performance indicators in recent years. An early reference cited by Thiry[14] is the 1994 paper by Kirk presented to SAVE International (at that time the Society of American Value Engineers). Thiry illustrates Kirk's quality model as a radar diagram comprising the following performance measures:

- Capital cost
- O&M cost
- Schedule
- Operational effectiveness
- Flexibility/expandability
- User comfort
- Site planning image
- Architecture image
- Community values
- Engineering performance
- Security/safety in operation
- Environmental

In July 2002 (*Architect's Journal*, 11 July 2002) the Construction Industry Council launched the Design Quality Indicator, which has significant similarities to Kirk. The indicators are grouped under three headings as follows.

- Functionality:
  - ☐ Use
  - ☐ Access
  - ☐ Space
- Build quality:
  - ☐ Performance
  - ☐ Engineering Systems
  - ☐ Construction

- Impact:
  - ☐ Form and materials
  - ☐ Internal environment
  - ☐ Urban and social integration
  - ☐ Character and innovation
- Additionally the topics of finance, time, environment and resources are dealt with separately.

The Construction Best Practice Panel (CBPP) performance criteria for benchmarking are:

- Construction cost
- Construction time
- Predicted design cost
- Predicted design time
- Defects
- Client satisfaction product
- Client satisfaction service
- Profitability
- Productivity
- Safety

While Kirk's model and the Construction Industry Council Design Quality Indicator focus on the product, the CBPP indicators focus on the process with the exception of the all-embracing 'client satisfaction product'.

This analysis has identified clearly that there is a close affinity between TQM and VM. QA can act within or outside of this but is process focused to deliver the product. Performance indicators in construction, which are an important ingredient in both TQM and VM, focus on product (e.g. Kirk and CIC DQIs) and process (e.g. CBPP). Both types of key performance indicators are vital for defining value systems.

8.5 A method for the discovery of the client's value system

Introduction

Value is defined above as a relationship between cost and quality. A methodology for deriving the client's value system that is capable of audit is a key prerequisite within an environment increasingly dominated by procurement criteria and systems that demand tender decisions based upon best value rather than lowest cost. Value should incorporate also the dimension of time since time and cost are linked in the majority of projects.

This section demonstrates through two exemplar case studies different approaches to defining and making explicit the client's value criteria.

Method 1: the time, cost and quality triangle

The time, cost and quality triangle is widely used in VM workshops as a tool to elicit from the client their value criteria. The value manager commonly asks for team consensus on the position of the dot. This indicates the team's relative value criteria in terms of the three variables time, cost and quality. The discussion commences with the client stating that all are important and therefore the dot should be in the centre. It is only after protracted discussion that the position of the dot tends to move. Figure 8.5 illustrates a time, cost and quality triangle for a proposed new court project. The dot indicates the result of a discussion in which the team agreed that quality is more important than cost and that time is not important in the context of a replacement for an existing court that is anticipated to have a life in excess of 100 years.

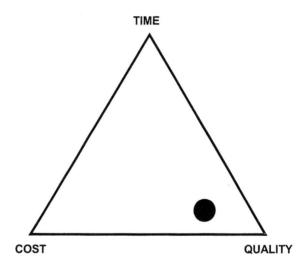

Fig. 8.5 Example time, cost and quality triangle.

The time, cost and quality triangle is a useful catalyst to discussion within a team but is inexact and often leads to lengthy debate particularly regarding the components and scaling of quality.

Time, cost and quality expounded

To understand value in more depth is fundamental if value is to be managed. At the information stage of a value management workshop it is necessary to reach consensus on the value system used by the corporate client body, and this value system has to be sufficiently overt for its meaning to be understood by the design team. Shillito & De Marle[15] offer the view that value is the primary force, a potential energy field that motivates human actions in respect of the objects they desire or need. Therefore the value of products or services can be represented by the presence of positive and negative features that attract or repel the customer. Understanding these gives the value system of the customer. This is a useful concept as it brings a much tighter relationship between value and quality as represented by the models described above.

A number of value engineering texts state that value denotes a relationship between function, cost and worth (Parker[16]) or function, cost and quality where value equals function plus quality divided by costs (Dell'Isola[17]). Functionally equivalent options to meet users' needs, desires and expectations, with varied cost, quality and worth attributes, are sought usually for the purpose of improving value. This concept can be summarised as 'the lowest cost to reliably provide the required functions or service at the desired time and place and with the essential quality' (Mudge[18]). Value is expressed in these texts as:

- *Exchange value* which relates to worth or the monetary sums for which products can be traded. This may be different from market value defined as the sale price of a product under the voluntary conditions of a willing buyer and a willing seller.
- *Esteem value* which relates to oneself or the monetary amount to be paid for functions of prestige, appearance and/or other non-quantifiable benefits. It may also refer to the monetary measure of the functions of a product that contribute its desirability (including obtaining for the sake of possession).
- *Use value* which relates to need or the life cycle cost considering user function only or those properties that accomplish a use by some work or service.

Fallon[19] states that the concept of value is encapsulated in the phrase 'an idea that relates objectives to one another and the cost of obtaining them'. Fallon also quotes from Adam Smith's *Wealth of Nations* on the paradox that extremely useful goods, such as water, have little or no exchange value whereas certain other goods of comparatively little use, such as diamonds, have great exchange value. Miles[20] states that the degree of value in any product depends on the effectiveness with which every usable idea, process, material and approach to the problem have been identified, studied and utilised. Value is therefore considered good if a product of lower or equivalent cost to its competitor contains a somewhat better combination of ideas, processes, materials and functions. Miles is also quoted by Fallon as having stated that poor value is a people problem, again relating back to the paradox that individuals come to quality with different knowledge and experience.

Financial management texts introduce the concept of stakeholder value. Scott[21] states that the principal goal of management is to maximise the level of sustainable growth in profitability and thereby enhance shareholder value, defined as the maximising of returns to those who have an ownership stake in the business. Porter[22] discusses value in the context of the economics of substitution, making the point that different buyers will evaluate substitution differently when faced with equal economic inducements for a functionally suitable equivalent product or service. The understanding of the buyer's value system in this situation is fundamental to the supplier's continued existence. Harvey[23] makes a similar point, stating that the value of a product or service is significant in economic terms only when a person is prepared to give up something in order to obtain it. An individual's value system will determine how much to give up and therefore different goods will have differing values. Value is measured in terms of the opportunity cost to the owner or purchaser.

In summary value is defined here as a relationship between cost, time and quality where quality is comprised of a number of variables determined from the knowledge and experience of an individual, or of individuals within a group, made explicit for the purposes of making choices between functionally suitable options. The value system made explicit is therefore the representation, at a fixed point in time, of a discrete range of variables against which all decisions affecting core business or a project can be audited.

Method 2: making explicit the variables of quality and value

On numerous occasions the authors have used the time, cost and quality triangle in VM studies to generate a debate about the relative importance of the parameters as a measure of value. However, it was thought that another tool that incorporated additional variables was required to more accurately define the client's value system. A review of the value management, quality and construction literature identified use or fitness for purpose, aesthetic or esteem and exchange or resale price as potential variables for inclusion. Quality was incorporated as internal and external quality. In addition, reflecting the current debate on the environment, this subject was considered a worthy addition. The aim was to bring the whole team into line or into focus with respect to a common view of the client's value system. Paired comparison was chosen as a team orientated technique to differentiate between the variables. The value manager asks the team to decide which of two variables is more important. For example, using the matrix below the question is asked, 'Which is more important, time or cost?' In the case below, cost was seen as more important and the letter B returned in the box. This is repeated for each pair of variables. The number of As Bs and Cs etc. is totalled to give an overall priority.

The example chosen here to explore the use of this approach is a visitor centre to be sited in an area of outstanding natural beauty. The team included project architects (two), engineer, surveyor, client representatives of different head office departments (three), the client's in-house facilities manager, the park ranger (employed by the client) and the local councillor.

Figure 8.6 illustrates the way in which the team understood the client's value system. In this case exchange, i.e. for how much the building could be sold, was not considered as a factor as the building will never be sold. Cost was seen as the least important factor when compared with the other factors. Time was also seen

210 Frameworks of Value

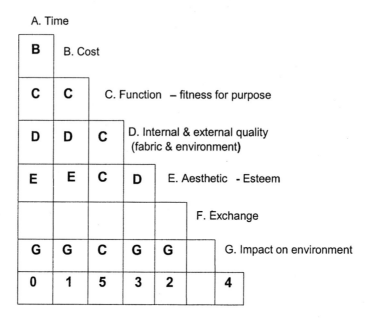

Fig. 8.6 First value system.

to be of little significance in relation to the others. The team therefore judged the relative importance, from highest to lowest, to be fitness for purpose, then the minimising of environmental impact, then quality and then aesthetic.

The result reflects the views of the whole team, client executive, client operations and design team. Disagreements, mostly centred on quality, were resolved by voting. It was felt by the value manager that the result did not reflect a consensus of views of the three groups and this was seen as a concern with this approach. The design team became obstinate on certain elements of the value system particularly in respect of environment. The architects were unable to distinguish between internal and external quality in terms of finishes, fittings, furnishings and space and the extra-over for aesthetic expression and esteem.

The value managers in a post study review considered the result obtained unsatisfactory. There was more disagreement within the team than was the case with the time, cost and quality triangle method. An analysis of the workshop proceedings revealed that the design team were attempting to influence vocally the client's value system rather than to learn from the debate. Further, something more than the earlier debate had to be done to address quality within the method.

Quality and value in relation to building design

In the analysis of architectural design, Vitruvius[24] stated in 100 BC that the value system for architecture depends on order, arrangement, eurythmy, symmetry, propriety and economy. Pena[25] states that definition of a project for the purposes of an architectural programme can be arranged under the headings of function, form, economy and time.

Duerk[26] states that design issues are processed through the filter of the values of the client, user and designer to yield goal statements about the qualities the design must have to ensure success. A goal is a concise statement of the designer's promise to the client about the quality of the design in relation to a particular issue. The quality is determined within the two extremes of the quality continuum and defined in terms of levels of performance. However, this has to be realised in context; for example, rooms in a Japanese house may be separated only by paper walls because the cultural rules of polite behaviour require that sounds from other rooms be thoroughly ignored and therefore privacy is assured because sounds are 'unheard'. Western culture has no such prohibition so one of the major complaints about hotels is that rooms are not private enough because the wall construction has not dampened sufficient sound. In this context the public/private continuum is relative.

Kirk[27] describes typical design objectives as being the aesthetic, esteem or image, the concept of the building and the way in which the building attracts attention to itself. Functional efficiency and flexibility is the degree to which the building is able to respond to the work process and flow of people, equipment and materials, and to be rearranged or expanded by the client to conform to revised processes and personnel changes with minimal disruption to existing building functions. Human performance is impacted by the physical and psychological comfort of the building as a place for working and living, supported by technical performance and how the building operates in terms of mechanical systems, electrical systems and industrial processes. Life cycle costs are described as the economic sequence of building in terms of initial capital investment and then long term operating costs. Good neighbour issues cover the impact on the community, energy conservation and security, addressing the degree to which the building can segregate sensitive functions from one another and prevent the entry of people to restricted areas. Kirk[28] demonstrates the use of weighting design objectives as a methodology for highlighting the relative degrees of importance, or value priorities, of the various design objectives.

Best & De Valence[29] highlight the complexity of quality by listing for illustrative purposes 15 factors that may be subjected to a quality continuum. Davies[30] describes sets of scales for setting occupant requirements and the rating of office buildings. These scales, published in a volume of over 300 pages, describe factors not unlike the quality continuum factors described above.

The conclusion is that while cost and time are reasonably straightforward the concept of quality is less so. This must lead to the question of how meaningful the time, cost and quality triangle really is when a client organisation is asked to reach a global conclusion with regard to quality. Duerk[31] gives a clue to an alternative approach by stating that design issues are processed through the filter of the values of the client. If value is comprised of cost, time and other factors summarised by many in the term quality then maybe it is more productive to examine the hidden factors of value and make them explicit.

In a traditionally procured construction project the client value system becomes established through a process of trial and error on the part of the designers. It evolves slowly over time as the design team present and re-present schemes that reflect their current understanding of the client's value system. With each iteration the designers take one step closer to full understanding. However, the newer procurement systems are not sympathetic to this slow iterative process. It is proposed here that the client's value system is made overt initially in a single operation, for later validation through a process of discovery using the value management techniques described below. In order to undertake this only client representatives are permitted to speak, with other participants as listeners.

For the client's values system to be meaningful the variables of time, cost and quality must be capable of description and measurement. The key to making the client's value system overt, and therefore auditable, lies in understanding the description of quality. To derive a measurable statement of quality it needs to be uncovered and made explicit. A synthesis of the previous quality and value debate denotes that project quality can be represented by environment, exchange, politics/popularity, flexibility, esteem and comfort each of which have their own continuum. The components of the full client value system therefore become:

- *Time*, from the present until the completion of the project, the point when the project ends and is absorbed back into the core client business. Time can be assessed on a continuum from where time is 'of the essence' to where time is 'at large'. The former means that were the project to be delivered even one day late it would be of no value, for example a contract for the supply of a satellite which is delivered late and misses a shuttle launch.
- *Capital cost (CAPEX)* are all costs associated with the capital costs of the project, measured on a continuum between the budget being considered tight and not able to be exceeded to there being flexibility in budgeting. In some situations the capital investment is subsumed within the operating cost and therefore the capital cost variable is omitted. This can occur, for example, where the cost of a building is rentalised within a total lease package, such as within a Private Finance Initiative project.
- *Operating cost (OPEX)* refers to all costs associated with the operations and maintenance implications of the completed project as it moves to an operational product within the client's core business. In the context of a building this includes facilities management which may be limited to maintenance, repairs, utilities, cleaning, insurance, caretaker and security but may be expanded to include the full operational backup such as catering, IT provision, photocopying, mail handling and other office services. The continuum is between OPEX having to be at a controlled absolute minimum to there being some flexibility in operating cost.

- *Environment* refers to the extent to which the project results in a sympathetic approach to the environment, measured by its local and global impact, its embodied energy, the energy consumed through use and other 'green' issues. The continuum is between maximum observance of Kyoto and Agenda 21 issues to indiscriminate sourcing policies and solving every problem by adding more power.
- *Exchange* or resale is the monetary value of the project. This may be viewed as assets on the balance sheet, the increase in share value, capitalised rental or how much the project would realise were it to be sold. The continuum is between requiring maximum return and return being of no consequence. If the physical asset is never to be sold, as in the case of the visitors centre example above, this item would be scored as zero in the value system equation.
- *Flexibility* represents the extent to which project parameters have to reflect a continually changing environment in the design. This value criterion is generally associated with changing technology or organisational processes or both. For example, medical practice is changing so rapidly that spaces in a hospital may need to accommodate a number of differing functions during the life of the building. The continuum is between being highly flexible and able to accommodate changing functions to being unlikely to change to any extent. If the project does not have to accommodate any flexibility this variable is scored as zero.
- *Esteem* is the extent to which the client wishes to commit resources for an aesthetic statement or portray the esteem of the organisation, internally and externally. The continuum is between needing to attract the admiration of the world to esteem being of no significance.
- *Comfort* is the physical and psychological comfort of the building as a place for working and living and how this will impact human performance. Comfort is measured on a continuum from the support of the business in purely utilitarian terms to a high degree of opulence.
- *Politics* is an external dimension that refers to the extent to which community, popularity and good neighbour issues are important to the client. The continuum ranges from requiring to be popular with the local community or electorate to having no concerns towards neighbours.

To derive a client's value system a pairs comparison exercise is undertaken using the matrix given in Fig. 8.7. Only the client representatives may speak during this process; the design team, contractors' representatives and any other stakeholder not a part of the client body must keep silent and listen, for this is the client's value system. Each box represents a question phrased 'which is more important to you...?' or 'would you be prepared to sacrifice...?'. Either way the letter inserted in the box represents whichever factor is the more important. For example, the question may be posed: 'are you prepared to spend more now to offset costs in the future?'. If the answer is: 'yes, I am prepared to spend more now to offset future costs' then obviously future costs are more important to you than capital costs and therefore the letter B is entered in the box. Conversely if

214 Frameworks of Value

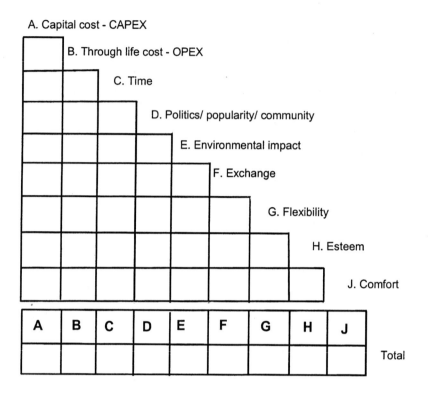

Fig. 8.7 Method 2.

the answer is: 'no, I must stay with the present budget even if it costs me more later', then future costs are less important to you than capital costs and therefore the letter A is entered in the box.

The number of times that A appears is entered in the total box and likewise for all of the other headings. The individual units of the value system can therefore be ranked to represents the overall client's value system. This may then be checked back with the client. The paired comparison method is useful when working with a number of client representatives as it allows discussion to occur at two levels when discussing two variables only and, on completion, discussion can confirm the final result of the exercise.

Workshops have found the paired comparison approach a satisfactory method of deriving a client's value system judged by the fact that clients generally agree with the summary when it is read back to them. For example, consider the sports centre shown in Fig. 8.8. The summary of the paired comparison questioning is:

■ It is of equal primary importance that the project is delivered on budget and that it maximises internal comfort and satisfaction for the users.

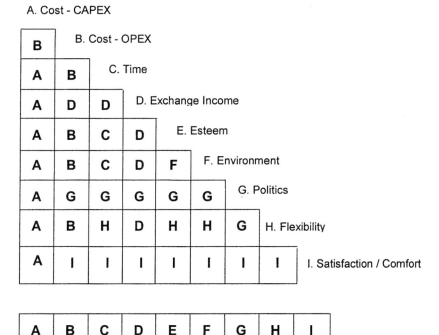

Fig. 8.8 An example client value system for a sports centre.

- It is also of high importance that the sports centre is popular within the local community.
- Finally it is important that operating costs are controlled and revenue opportunities maximised.
- Issues such as the timeliness of delivery of the project, environmental impact, flexibility in use and architectural esteem are of lesser importance compared with the above.

8.6 Conclusion

This chapter concludes the section on value concepts by discussing the relationship between the client's value system and the quality systems of total quality management and quality assurance. The former is culture driven whereas the latter is process driven and can be undertaken without reference to the former. An association is recognised between the client's value system for undertaking a value management study and the fundamental basis for total quality management. This recognition comes from the need to capture the value culture of an organisation explicitly and reflect this in the definition of the product and through the effective and efficient delivery of the process. Quality is a construct of comparability and if the client does not establish explicitly the value criteria but transfers that responsibility to the design team, the latter will overlay their own value criteria on the project. However, it is argued that only the client representatives should speak and other members of the team should participate as listeners. Having explicitly expressed their value criteria as a matrix displayed for all to see, it can be used subsequently in establishing the strategic function logic diagram, which describes why an investment takes place, and also the value engineering element component diagram, which incorporates the CVS into element function priorities. This can be transposed by the designers into detailed product specifications.

8.7 References

1. Burt, M. E. (1975) *A Survey of Quality and Value in Building*. Building Research Establishment, Garston, Watford, Herts.
2. Bicheno, J. (2002) *The Quality 75: Towards Six Sigma Performance in Service and Manufacturing*. PICSIE Books, Buckingham.
3. Juran, J. M. & Gryna, F. M. (1988) *Juran's Quality Control Handbook,* 4th edn. McGraw-Hill, New York.
4. Vorley, G. (1998) *Quality Management: Principles and Techniques*, 3rd edn. Quality Management and Training (Publications) Ltd, Guildford.
5. Pirsig, R. M. (1991) *Zen and the Art of Motorcycle Maintenance*. Vintage, London.
6, 7. Bicheno, J. (2000) *The Lean Toolbox*, 2nd edn. PICSIE Books, Buckingham.
8, 9. Vorley, G. (1998) *Quality Management: Principles and Techniques*, 3rd edn. Quality Management and Training (Publications) Ltd, Guildford.
10. Deming, W. E. (1986) *Out of the Crisis*. MIT Press, Cambridge, Mass.
11. Juran, J. M. & Gryna, F. M. (1988) *Juran's Quality Control Handbook*, 4th edn. McGraw-Hill, New York.
12. Breyfogle, Forrest W. III (1999) *Implementing Six Sigma: Smarter Solutions Using Statistical Methods*. 2nd edn. Wiley, Hoboken, NJ.
13. Kaplan, R. S. & Norton, D. P. (1992) *The Balanced Scorecard*. Harvard Business Review January/February, Harvard Business School, Boston, Mass.
14. Thiry, M. (1997) *Value Management Practice*. PMI Publications, Philadelphia.
15. Shillito, M. L. & De Marle, D. J. (1992) *Value: Its Measurement, Design and Management*. Wiley, Chichester.
16. Parker, D. E. (1977) *Value Engineering Theory*. Lawrence D. Miles Value Foundation, Washington D. C.

17. Dell'Isola, A. (1997) *Value Engineering: Practical Applications for Design, Construction, Maintenance and Operations*. R. S. Means Co, Kingston, Mass.
18. Mudge, A. E. (1989) *Value Engineering: A Systematic Approach*. J. Pohl Associates, Pittsburgh.
19. Fallon, C. (1980) *Value Analysis*, 2nd edn. Lawrence D. Miles Value Foundation, Washington D. C.
20. Miles, L. D. (1989) *Techniques of Value Analysis and Engineering*, 3rd edn. Lawrence D. Miles Value Foundation, Washington D. C.
21. Scott, M. C. (1998) *Value Drivers: The Manager's Guide to Driving Corporate Value Creation*. Wiley, Chichester.
22. Porter, M. E. (1985) *Competitive Advantage: Creating and Sustaining Superior Performance*. The Free Press, New York.
23. Harvey, J. (1984) *Modern Economics*, 4th edn. Macmillan, Basingstoke.
24. Morgan M. H. (1960) *Vitruvius: The Ten Books on Architecture*, Dover Publications Inc., New York.
25. Pena, W., Kelly, K. & Parshall, S. (1987) *Problem Seeking*, 3rd edn. AIA Press, Washington.
26. Duerk, D. P. (1993) *Architectural Programming: Information Management for Design*. Wiley, Chichester.
27, 28. Kirk, S. J. & Spreckelmeyer, K. F. (1993) *Enhancing Value in Design Decisions*. S. J. Kirk, Detroit, USA.
29. Best, R. & De Valence, G. (1999) *Building In Value: Pre-Design Issues*. Edward Arnold, London.
30. Davies, G., Gray, J. & Sinclair, D. (1993) *Scales for Setting Occupant Requirements and Rating Buildings*. International Centre for Facilities, Ottawa, Canada.
31. Duerk, D. P. (1993) *Architectural Programming: Information Management for Design*. Wiley, Chichester.

Part 3 The Future of Value Management

Value management is a process of eliciting explicitly, appraising and delivering the functional benefits and requirements of a client throughout the project value chain in a manner consistent with the client's value system. It achieves this by a series of interventions, at either project programme or individual project level, that take account of competing value systems that impinge on the strategic and tactical delivery of projects.

Part 3 comprises two chapters. Chapter 9, *Professionalism and ethics within value management*, focuses on VM as a management methodology, exploring its continued development within the context of the industrial structure of the UK. The chapter sets out the argument that VM is currently a methodology which places it within the 'craft/artisan' structure of the UK's industrial framework, where predominantly skills gained from experience following a period of non-degree level prescribed training are important for service delivery. The chapter reviews and defines occupations and professions, discusses the institutional framework for the certification of value management practitioners and concludes that a more extended service is required to move VM along a continuum of professionalism towards becoming a professional service more akin to management consultancy. The chapter considers the relationship between codes of conduct which are important for defining professions, and qualifying associations, the regulators and differentiators of educational and practice standards for a profession.

Chapter 10, *The future of value management*, considers the shape of an enhanced value management service within the major components of the orientation and diagnostic phase, the workshop phase and the implementation phase. Each phase is detailed and commented upon. The chapter also looks at a VM service that can be utilised as part of the government's best value initiative. The chapter concludes with three contrasting scenarios that tentatively map out future development paths for VM based on an extrapolation of current trends in the UK construction industry.

9 Professionalism and Ethics within Value Management

9.1 Introduction

This chapter explores some of the developmental issues now facing value management as a methodology. It builds on earlier work undertaken by the authors and on earlier chapters in this book[1,2,3,4]. It looks at the knowledge base of VM and how this translates into the practice setting. It discusses the market settings within which practice takes place and explores occupations, professions and different models of professions. Finally, the chapter reviews the implications of this analysis for value management.

9.2 The value management knowledge base: founded on theory?

Historically value analysis, the forerunner of value management through the work of Miles, was founded on the technique of functional analysis, with the structure for the process of applying the technique provided by the job plan. Functional analysis, it can be argued, is the only characteristic distinguishing value management from other management philosophies and approaches. Miles proposed the technique and the process for delivering VM in a manufacturing context.

The European standard for value management defines it as a style of management[5]. It notes its evolution from value analysis on products in the 1940s into services, projects and administrative procedures. It also notes that other methods and management techniques also based upon the concepts of value and function have developed contemporaneously, such as, design to cost and functional performance specification. The European and British Standard BS EN 12973:2000 identifies certain elements as essential for a successful VM study, illustrated in Fig. 9.1.

The standard indicates that the goal of value management is to reconcile differences in view between stakeholders, internal customers and external customers as to what constitutes value and by so doing enable an organisation to achieve the greatest movement towards its stated objectives using the minimum of resources. The Value for Europe Training and Certification System Manual[6],

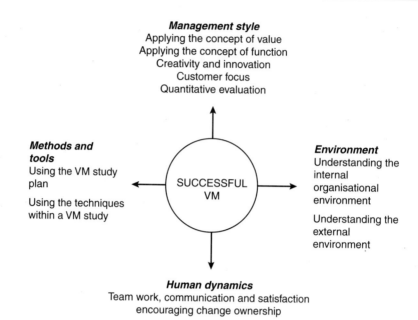

Fig. 9.1 The essential ingredients of a successful VM study.

supporting the deployment of the European Standard through training courses, identifies a taxonomy of learning objectives for value management and associated training requirements. The former is set out in Fig. 9.2 and the latter in Table 9.1.

The SAVE International standard adopts the term value methodology[10], which is abbreviated in the USA to VM and has the same meaning as the European term 'value management'. It highlights the fact that value methodology includes the processes known as value analysis, value engineering, value management, value control, value improvement and value assurance. The standard is all embracing. While it does not define the methodology per se, the SAVE standard embraces an approved job plan, a body of knowledge, typical profiles of value managers and value specialists, and the duties of a value organisation. The key components of the SAVE International body of knowledge are set out in Table 9.1.

The New South Wales *Total Asset Management Manual* (TAM2000) defines value management as a structured, analytical process for developing innovative, holistic solutions to complex problems[11]. TAM2000 highlights the underlying principle that there is always more than one way to achieve project objectives and that examination of the alternatives will produce the most acceptable conclusion. It recognises that at the core of the value management process is the analysis of functions from the point of view of the system as a whole (including the relationship or cost impact of design decisions on the project and/or scheme operation) and it is this aspect in particular that distinguishes value management from other methods of improving value.

Category \ Level of attainment	Information (or knowledge)	Practice (or skill)	Attitude	'Practice' archetype
Low (L)	Awareness	Participate by using theory	Accept	Noviciate (trainee)
Medium (M)	Knowing memorising recalling judging	Synthesise assimilate disassemble evaluate apply	Behave	Acolyte (artisan)
High (H)	Understand comprehend translate infer extrapolate	Mastery	Advocate promote convince	Master (professional)

Fig. 9.2 A taxonomy of learning objectives for VM.
Source: adapted from (1997) *Value for Europe: Training and Certification System Manual*[7].

The benchmarking study, other research conducted by the authors and a brief review of the European and American standards provide sufficient empirical evidence to indicate that the knowledge base of VM, in contrast to that proposed by Miles, is ill defined. Many of the tools, techniques, models and methods can be found within other disciplines. In addition there have been attempts recently to broaden the knowledge base of VM to include concepts from operational research, such as objectives hierarchies. A fundamental operating domain of value management during the workshop phase also involves the management of teams, the domain of psychology, organisational psychology and organisational behaviour.

This brief analysis has highlighted that there is a theoretical underpinning to value management but that, with the exception of function analysis, its principles are derived from other disciplines. Furthermore, in some countries value management has been dominated by practitioner. Additionally, comparing countries internationally value management has taken root in academia in different ways, embedding itself much more in the higher education curriculum in some countries than in others.

The diversity of the knowledge base, tools and techniques is a strength and a weakness. The strength comes from the richness of tools, techniques and per-

Table 9.1 The value management knowledge base.

	European Value Management		SAVE International		Comments
	Concepts	Categories	Concepts	Categories	
Basic course (minimum 3 days)	Understand human dynamics	Understand essentials of individual and group work, including the design and work of a VM team	Fundamental value concepts	VM history and characteristics	Value for Europe places the following within the scope of VM tools and techniques: value analysis, function analysis, function-cost, pareto and creativity techniques, design to cost, functional performance specification QFD, reengineering, simultaneous engineering, quality management, benchmarking; cost benefit analysis, risk analysis, HAZOP, life cycle cost, marketing research, FMECA, evaluation methods, DFMA, cost modelling, multi-criteria analysis, design of experiments, statistical analysis, etc.
	Understand the design and importance of a communication network with customers	Understand and accept the value culture, including knowing criteria for successful communication, the organisational process of VM and the importance of building and maintaining customer relationships		Job plans	
	Participate in VM project initiation	Understand the objectives of the VM project, including describing customer needs, participating in VM project planning and strategy building		Basics of function analysis	
	Know and apply value methods	Know European and national practices and understand the methods and tools of VM, including project support systems		Function/cost	

	Know how to contribute to a VM project	Participate in using VM methods. Know how to transmit appropriate information to the team, including how to comply with a programme and report problems		Function/worth
	Know how to participate in developing a value culture	Know how to report results and successes, including enhancing experience with other people's results. Know how to improve personal VM knowledge and competence and promote a value culture		Creativity and creative processes
Advanced course (minimum 3 days)	Understand and master human dynamics	Understand how to set up and develop a VM team, understand group situations and master the management of a VM team, including the external team environment	Concept application	Defining functions
	Apply communication with internal and external customers	Promote projects and proposals to management, decision makers and all employees. Recognise the involvement of internal and external customers and adopt good communication during the project		Classifying functions

Table 9.1 *continues*

Table 9.1 contd.

European Value Management		SAVE International		Comments
Concepts	Categories	Concepts	Categories	
Understand how to set up a project	Understand how to define a project and its start up, including building a project plan and strategy and contributing to a quality plan. Know the customer and their needs and be able to measure the effectiveness of a VM project		Allocating costs to functions	
Apply and develop expertise in the use of VM methods	Know European and national practice, be able to apply methods and tools, to select, apply and adapt them appropriately within VM. Know how to develop new VM concepts and tools and promote continuous training and improvement		Determining the value index	
Know how to manage interface with non-VM projects and activities	Know about the need to co-ordinate multiple VM projects within a wider programme, different types and goals of strategic, organisational and operational projects, including competing aspects, trade offs and external influences		Function modelling.	

Understand how to interact with business objectives	Understand how to act consistently within corporate strategy. Know how to build and implement VM policy		Financial analysis
Know and promote continuous improvement based on experience	Learn from best practice, from failures and how to improve expertise		Contracting for value methodology
Master and control the execution of a VM project	Apply and adapt a project plan. Master VM project and facilitation and know how to control a VM project, including ensuring its quality		Break-even analysis
			Life cycle cost analysis
Understand and promote the development of a value culture	Understand the importance of and how to promote a value culture outside and within the organisation		
		Management of value	Value studies
			Organising and implementing VM programmes
			Management's role
			Value team compositions
			Team leader skills
			Managing VM teams and studies
		Value analysis education	Basic VA education – training modules
			Related programmes

Sources: Adapted from *Value for Europe: Training and Certification System Manual* (1997)[8] and *SAVE International Value Methodology Standrd* (October 1998)[9].

spectives that can be drawn on to solve client value problems. The weakness is the lack of clarity in defining exactly what value management is. This lack of clarity in the UK has the potential to lead to interdisciplinary and interprofessional 'turf warfare' over who 'owns' value management. This is taken up in a later section.

The individual practitioner knowledge and skill base

The European standard recognises three types of value management study:

- Strategic
- Organisational
- Operational (project orientated)

Earlier chapters have also identified different study styles, each of which will fall into the the types identified by the European standard BS EN 12973: 2000.

In the authors' view value management encompasses a generic process comprising the orientation and diagnostic phase, the workshop phase and the implementation phase. An earlier chapter has also highlighted that there are different facilitation styles for the workshop phase:

(1) Facilitate workshop(s) only.
(2) No preparation; facilitate workshop(s) and produce a report.
(3) Some preparation and orientation; facilitate workshop(s) and produce a report.
(4) Full preparation, orientation and diagnosis; facilitate a workshop(s) and produce a report.

Tables 9.2 and 9.3 set out the VM process (study plans) as identified in the European standard, the SAVE International standard, the International VM Benchmarking Study and the New South Wales TAM2000 Manual, for ease of comparison.

Earlier chapters have also argued that a value manager requires the following skills and knowledge base to undertake studies:

- Strategic and tactical skills to plan, organise and deliver a study to meet preset objectives and to adapt the process to meet evolving situations as they arise.
- Problem structuring and information synthesis throughout each successive phase of a VM study.
- Experience gained from a variety of different project and organisational settings.
- Competence in a variety of management skills related to human dynamics and the management of teams.
- Leadership qualities.
- The ability to use functional analysis.
- The ability to adopt and adapt other techniques to the value management process.

- The ability to structure and synthesise information and set workshop agendas to achieve deliverable outcomes from the study.
- The ability to manage the process of the workshop including:
    - ☐ Dealing with any hidden agendas of participants.
    - ☐ Recognising individuals and their contribution.
    - ☐ Questioning and summarising.
    - ☐ Providing direction and a sense of common purpose.
    - ☐ Sensing interpersonal relationships within the team.
    - ☐ Sensing the climate of the workshop.
    - ☐ Synthesising and integrating information during the workshop.
    - ☐ Intervening in the workshop process when necessary.
- The ability to communicate verbally and producing the VM report.

Table 9.2 VM job/work/study plans compared.

EN 12973:2000 Value Management Study Plan	EN 12973:2000 Value Analysis Work Plan		SAVE International Standard: Value Methodology job plan	
Activities	Stage	Activities	Stage	Activities
Define objectives in relation to VM policy and programme	**Preliminary phases (0)**	Project outline	**Pre-study**	Collect user/customer attitudes
		Feasibility investigation, risk analysis		Complete data file
		Profitability investigation, what is at stake		Determine evaluation factors
		Decision maker and VA project leader selection		Scope the study
Identify methods and supporting processes to achieve objectives	**Project definition (1)**	The subject		Build data models
		Framework for the study		Determine team composition
		Premises of the data about the problem		
		Marketing objectives		
		General objectives		
		What is at stake		
		Resources		
		Participants		
		Preliminary risk analysis		
Select team, define their roles and assess competencies	**Planning (2)**	Constituting a work team		

Table 9.2 *continues*

Table 9.2 contd.

EN 12973:2000 Value Management Study Plan	EN 12973:2000 Value Analysis Work Plan		SAVE International Standard: Value Methodology job plan	
Activities	Stage	Activities	Stage	Activities
Provide training as necessary to apply methods and work as team		Working out an initial time schedule		
		Agreeing venue for work		
	Gathering comprehensive data about the study (3)	Information gathering	Value study	Information phase
		Detailed market survey		Complete data package
		Miscellaneous other information		Modify scope
Identify essential functions to meet objectives completely	Function analysis, cost analysis, detailed objectives (4)	Expression of need and function analysis		**Function analysis phase**
Agree how to measure changes in performance and use of resources		Cost analysis and function cost		Identify and classify functions
Set targets for performance and use of resources		Fixing detailed objectives and evaluation criteria		Develop function models
				Establish function worth
				Cost functions
				Establish value index
				Select functions for study
Apply methods and supporting processes, identify innovative ways to achieve targets	Gathering and creation of solution ideas (5)	Gathering existing ideas		**Creative phase**
		Creation of new ideas		Create quantity of ideas by function
		Critical analysis		**Evaluation phase**
Select and develop proposals for implementation at appropriate predetermined points in the project	Evaluation of solution ideas (6)	Evaluation of ideas		Rank and rate alternative ideas

EN 12973:2000 Value Management Study Plan	EN 12973:2000 Value Analysis Work Plan		SAVE International Standard: Value Methodology job plan	
Activities	Stage	Activities	Stage	Activities
		Choice of what is to be developed		Select ideas for development
		Work programmes for development		**Development phase**
	Development of global proposals (7)	Studies, tests, industrial development		Conduct benefit analysis
		Follow-up co-ordination		Complete technical data package
		Evaluation of solutions: • Qualitatively • Economically • Risk analysis		Create implementation plan
Validate proposals and implement them	**Presentation of proposals (8)**	Selection of proposed solutions		Prepare final proposals
		Developing implementation programmes		Present oral report
		Organising comprehensive data on proposals		Prepare written report
		Obtain decision from decision maker		
		Keep VA team informed; dismiss or place on standby		Obtain commitment to implementation
	Implementation (9)	Support implementation: • Follow-up • Assist to correct deviations	**Post study**	Complete changes
		Organise VA team to tackle unexpected problems		Implement changes
Monitor progress		Assess actual results with those expected		Monitor status
		Disseminate results to: • VA team • Experts concerned • Wider organisation		
Monitor and measure outcomes, compare achievements with targets, provide feedback for continuous improvement		Collect information on field experience		

Table 9.3 VM job/work/study plans compared.

International VM Benchmarking Study		TAM 2000	
Stage	Activities	Stage	Activities
Pre-study information	Information gathering	Information phase (pre-workshop)	Scope
	Information synthesis		Canvas issues and concerns
	Agenda production		Establish VM study objectives
			Identify and prepare background material
			Logistics
			Formation of value study team and technical specialists
			Establish study timetable
			Identify and gain commitment of stakeholders
			Nominate, invite and brief participants
			Arrange venue
			Brief presenters for workshop component of information phase
			Distribute consolidated background material
Workshop phase	**Information**	Workshop	**Information phase**
	Presentation and team building		Confirm value study objectives
	Information gathering		Scheme and project overview
	Information synthesis		Project assumptions
	Function logic diagramming		Project imperatives in terms of costs and funding
	Function analysis		Project imperatives in terms of time and other criteria
	Process analysis		Key issues and concerns
	Target functions		Presentations by key participants
	Creativity		**Function analysis phase**
	Brainstorming ideas		Identify and analyse functions
	Evaluation		What does it do?
	First level sort		What must it do?
	Refined sort		What does the function cost?
	Select ideas for development		What is the function worth?
	Development		Function hierarchies and FAST
	Develop ideas		**Ideas/options phase**
	Action planning		How else can required functions be performed?
	Presentation to sponsors and senior managers		What else will perform the function?
	Plan for implementation		What will the alternative cost?
	Prepare action plan		**Evaluation**
	Sign-off by participants		Option evaluation: criteria and ranking

International VM Benchmarking Study		TAM 2000	
Stage	**Activities**	**Stage**	**Activities**
			Option development
			Develop best ideas
			Action plan
			List activities to be carried out
			Identify people responsible for those activities
			Indicate time frame for each activity
			Specify finalisation date
			Appoint action plan coordinators and also nominees for each task
			Identify follow-up sessions for action plan nominees one month after value study workshop
Workshop report	Circulate report	**Analysis and reporting**	Interim report
			Final report
Implementation	Feedback workshop		
	Prepare final action plan		
	Sign-off by participants		

These skills are generally high level cognitive problem solving and 'doing' skills with their origins in the integration at individual practitioner level of education, experience, training and effective interpersonal behaviour. However, the extent to which all or some are used depends on the facilitation style adopted for the workshop phase. In addition, the value manager must be trained in the appropriate use and tailoring of the VM process and the relevant tools and techniques.

The one technique differentiating VM from other management methods is function analysis. Once learned this skill can be used in a variety of different situations as a natural language problem mapping tool. It can be used mechanically. Higher level cognitive functioning comes into play for the practitioner when the technique is used and adapted flexibly to meet the requirements of different situations. It is difficult to teach this; it only comes from practice, experience and reflecting on lessons learnt. The method, tools and techniques provide the practitioner with a link back to the knowledge base underpinning the service, which as argued above, is both diverse and rich.

The benchmarking study highlighted that knowledgeable, regular procuring clients have allegiances to particular value managers rather than the employing consultancy. This indicates clearly that not everyone is capable of being a good value manager and that they are potentially a scarce resource in the marketplace.

9.3 The influence of the marketplace on the provision of value management services

It has been argued elsewhere that clients would be a primary mechanism for diffusing value management practice and procedure[12]. The same study also highlighted experience from North America suggesting it would take about a decade for VM to become embedded into mainstream regular practice within the UK construction industry. In hindsight that prediction was uncannily accurate, with the Egan Report endorsing VM as an important method for use in construction[13]. Furthermore, the benchmarking study conducted in 1998 continued to argue the importance of commissioning clients for diffusing VM practice.

Large, knowledgeable and regular procuring clients were highlighted as using VM on a recurring basis. Often they will have highly developed methods for choosing VM practitioners. The value methodology is also likely to be embedded in a project delivery manual as a formal review procedure. However, complementary research on VM involving the authors revealed that small and medium sized less knowledgeable clients are often difficult to identify. Diffusion to them of VM practice and procedures is therefore hampered[14]. They are likely to be irregular users of the service. The benchmarking study also confirmed that the conclusions on large, regular procuring clients for the UK were replicated in North America and Australia. The clear inference from this is that a value management service offered by practitioners is under the patronage of large corporate, experienced, regular procuring clients. A flow or trickle down effect to irregular procuring clients is possible but difficult.

Equally, within regular procuring 'clients' of the service organisational units will exist staffed by VM practitioners who commission, monitor or undertake studies in their own right. These practitioners provide an internal service, often within the scope of an investment division or as a separate value audit function where a programme of projects is being delivered. The internal service can be broken down into:

(1) Setting up and monitoring value programmes but with no internal delivery.
(2) As 1 above but also including internal delivery.
(3) As 2 above but also encompassing external delivery.

Points 1 and 2 above are similar to the situation found in manufacturing companies where VM is embedded into their ongoing organisational processes as one of a number of management systems for improving products.

A typology of VM practice

The benchmarking study identified that the majority of VM practitioners see the provision of value management services as one of short intensive interventions at specific value opportunity points in the life of a project. Also, a clear message from the benchmarking study was that clients identify certain value managers as being suitable for particular types of projects, hence, their choice of practitioner is based upon individuals rather than the employing consultancy practice. Observations by the authors in conducting action research studies, consultancies and training activities continue to confirm this finding. The benchmarking study identified a typology of VM practitioners and organisations, indicating different tensions and pressures stemming from the market place. This was enhanced subsequently[15] and is refined further here to reflect observations on practice since previous publications. The term 'practice' is used here to refer to the practitioner roles operating under a series of different market structures, either internally within an organisation or externally in the market place.

The *Type 1* practice is the sole trader (or perhaps two at most) operating within the confines of a pure 'cottage industry' role. The practitioner is able to compete very effectively on price with a low overhead structure. Their main pressures in the market place stem from the need to maintain a volume of work. Resourcing can become an issue when work expands. Consequently the Type 1 practitioner can operate within a network of other VM practitioners to offset the problem.

The *Type 2* practice is also a sole trader with similarities to the first type but in working association with a larger practice from within their premises but not directly employed by them. They have the advantage of marketing under an umbrella consultancy organisation and can be sub-contracted in by the overarching organisation to offer or use VM on its own projects.

A *Type 3* practice is a consultancy offering VM as their core service. They have volume from turnover and offer a unique specialist service permitting fees to be charged with the intention of sustaining the business over the longer term. This type may be susceptible to internal tensions because clients prefer particular VM practitioners to others from the same consultancy organisation. There is a centrifugal tendency within this type of organisation for spin-off competing Type 1 practices to emerge, taking their clientele with them. The strong internal tension may become too great for the Type 3 to retain internal cohesion into the longer term with the attendant problems of resourcing and growth. The Type 3 practice has two sub-types. *Type 3a* is a stand-alone consultancy. *Type 3b* is a stand-alone consultancy that is a subsidiary of a larger corporate structure, perhaps a large consultancy group.

The *Type 4* practice operates as an internal VM function within a large corporate client, for example a utility organisation or a corporate client with programmes of projects. The Type 4 practice operates under internal and not external market structures and practitioners are paid as full-time staff by the organisation. The Type 4 practice comprises three sub-types. *Type 4a* is essentially a Type 1 practice but where the practitioner is only responsible for managing VM programmes and not practising. They will sub-contract in from the external marketplace those offering VM services under practice Types 1, 2, 3 and 5. Depending on their organisation's competitive situation and industry sector they may also commission Type 4c practices. The *Type 4b* practice has similarities to Type 3 except that their focus is to provide an internal VM service only. The primary pressure here is internal volume of work to justify a separate organisational unit. Internal charging mechanisms for the service also raise difficulties. The *Type 4c* practice can operate internally and offer external services in VM to other clients. The primary pressure and tensions here are first as to where priorities lie, internally or externally, and where the trade-off occurs when conflicting priorities arise. Second, offering an external VM service brings the Type 4c into direct competition with 1, 2, 3, 5 and potentially other Type 4c's. Again, the problem facing this type of practice is one of sustaining full-time employment from VM to the satisfaction of the employing organisation.

The *Type 5 practice* occurs where value management is not the sole activity of the practitioner(s). The advantage here is that the organisation and the individual are not reliant on VM to sustain business income. The main disadvantages lie in being less able to demonstrate commitment to VM to potential clients or keep abreast of VM current practice and issues, and the inability to create a wide enough portfolio of studies to convince a diversity of clients that the practice can operate across a range of industrial sectors. To some extent, depending on the parent organisation, drawing on internal expertise can offset this difficulty.

The UK construction industry now has examples of all five VM practice types in existence. There are a series of implications stemming from the foregoing for the continued professional development of value management in the UK, and these are the subject of the next section.

9.4 Value management: methodology, occupation or profession?

Earlier chapters have argued that value management is a methodology and a mechanism for undertaking studies within the project value chain or other areas where value systems need to coalesce. The primary technique upon which the methodology was founded is that of function analysis. There are a number of different value management institutes internationally proffering to represent practitioner interests although it is also recognised that the founding institute is the Society of American Value Engineers, now retitled SAVE International. In the UK the Institute of Value Management represents UK interests within a wider European perspective. An earlier section in this chapter has also identified different settings within which a practice occurs through the exploration of a typology of practice.

This section addresses whether value management is just a methodology around which institutional and vested interests are coming together; whether it is an occupation in its own right; or whether it has taken or is taking on the mantle of becoming a new profession where normal ethical considerations are a critical element.

As a starting point for the analysis it is worth restating that the authors see value management as encompassing three phases:

- *The orientation and diagnostic phase.* The value management practitioner (the value manager) orientates him/herself to the value problem and diagnoses the appropriate structure for a study, the structure of the workshop phase and how outcomes are to be implemented on completion of the study. The value manager has intentionally defined the parameters for a successful study.
- *The workshop phase.* This is the means by which value systems come together, coalesce and introduce change and improvements to move things forward. The value manager is both a process manager and a facilitator during this phase, deciding tactically how the workshop process will achieve the objectives for the study set in the diagnostic phase. The strategy for the workshop phase revolves around its role within the overall study context. The tactics are about formulating the agenda, deciding on the tools and techniques to be used and adjusting these as the workshop process unfolds, and deciding on team working procedures.
- *The implementation phase.* This is the final process in the study and defines its ultimate success: the extent to which the costs and benefits have been realised through implemented solutions.

An earlier chapter has also talked about different study and facilitation styles. These reflect the extent to which the value manager approaches, designs and implements the process in a different way.

Techniques, methodologies, occupations and professions

Value analysis, as a methodology, used the technique of functional analysis and the process of the job plan to introduce improvements into manufactured products[16]. Value analysis evolved through political manoeuvrings: witness the emergence of the term 'value engineering', and developments in other areas of industry, such as construction and the service sectors, subsequently leading to the emergence of 'value management' as the broad conceptual term. As part of this developmental journey various institutes have formed internationally to capture it as their own territory. Equally, other institutes have argued that it is part of a wider set of skills which can be encompassed within existing occupational domains and professional groups. This tussle and turf warfare has all the hallmarks of the professionalisation process characteristic of the historical development of many present day professions. Where does value management currently sit in the process of professionalisation? In order to assess this it is necessary to define terms such as 'occupations', 'professions', and 'professionalisation'. Building on the work commenced in the benchmarking study, this section explores the following in reaching a series of conclusions about value management and its continued development:

- The defining characteristics of occupations and professions.
- The collective knowledge base of VM and whether this is founded on theory.
- The application of the theoretical knowledge base by practitioners.
- Education and training.
- Certification.
- Codes of Conduct, Standards, and ethical behaviour.

The next section looks at value management in the occupational structure of the UK.

Occupations

An earlier section identified a typology of VM practice, indicating that it can range from practitioners that may undertake it intermittently to those that practice it as a full time job. An earlier section also identified the knowledge and skill base of VM. The current section reviews material on the nature of jobs and occupations, setting the parameters for ascertaining whether VM has the capability of becoming a recognised occupation in the industrial fabric of the UK construction industry.

Occupations define and constitute a major focus in the life of an adult and are derived from the division of labour in society[17]. Occupations focus on what people do in their jobs which can be further categorised into dimensions covering[18]:

- The type of materials and tools used.
- The nature of the working environment.
- The skills required.
- The level of skills.

The status in employment approach adopted by the UK Office for National Statistics recognises those that are:

- Self-employed with employees (proprietors).
- Self-employed without employees (independent professionals and artisans).
- Employees with managerial status (managers).
- Employees with foreman or supervisory status (supervisors).
- Other employees (other workers).

Sociologists are interested in the way that societies are stratified socially and economically based on income, education and qualifications. The UK

Government's Standard Occupational Classification (SOC) relates types of occupational skill and level of occupational skill with particular job titles. Table 9.4, derived from the SOC, relates qualifications, training and experience to occupational groups with examples quoted from construction. Interestingly, supporting information to the table indicates that architects and civil engineers fall into the 'professional' grouping whereas quantity surveyors fall into the associate professional grouping. Management consultants are viewed as falling into the 'professional' grouping.

The UK Government currently utilises two socioeconomic classification systems. First, the Registrar General's Social Classes (RGSC) classifies occupations and job holders on an ordinal scale of social class and recognises six groupings including Professional (I), Intermediate (II – incorporating managers) and Skilled non-manual (IIINM). Second, the Socioeconomic Grades (SEG) does not rank occupations/job classifications but acknowledges managers, proprietors and self-employed workers as distinct groups within a major classification system. The SEG also takes account of the number of employees in an employer's establishment. Finally, the 'census matrix' relates occupational group to status in employment and whether a self-employed person has employees working for them in order to assign jobholders into categories. Implicitly the census matrix makes the link between employment characteristics and position in society as measured by social class. To summarise, there is a direct link between a person's job, official occupational classification and social status.

The United States Department of Labour, Bureau of Labour Statistics on Standard Occupational Groupings does not register value engineer, value manager or value analyst among its recognised job titles. The US SOC does register project managers under the generic title of construction managers, defining the project manager as one who manages, coordinates and supervises the construction process. Equally the SOC in the UK omits reference to value engineers or value managers as it does to project managers, although production managers is a recognised job title. Possible reasons for the lack of recognition of the 'value' related job title could be, first, that the title is too new to register as an occupational category. Certainly in the USA the activity of value engineering has been around for some 40 years, though in the UK for a much shorter period. Second, as a job activity it is assumed to occur within the remit of other occupational groupings such as engineers, management consultants or other design professionals. Third, the activities performed within value management and value engineering are insufficiently defined to create a definitive boundary for it to be recognised as a separate occupation.

The next section discusses value management in the context of occupations and professions.

The nature of professions

In his study of quantity surveying Male identified that professions are about power, status and money[20]. Professions have also been argued to be a special

Table 9.4 General nature of qualifications, training and experience for occupations in SOC major groups.

Major group	General nature of qualifications, training and experience for occupations in the major group	Typical examples in construction
Managers and administrators	A significant amount of knowledge and experience of the production processes, administrative procedures or service requirements associated with the efficient functioning of organisations and businesses	Managers in building contracting; clerks of works
Professional occupations	A degree or equivalent qualification, with some occupations requiring post-graduate qualifications and/or a formal period of experience related training	Civil engineers; architects; building, land, mining and general practice surveyors; management consultants
Associate professional and technical occupations	An associated high level vocational qualification, often involving a substantial period of full time training or further study. Some additional task related training is usuallly provided through a formal period of induction	Architectural, building and civil engineering technicians; quantity surveyors and quantity surveyors' assistants; estimators and valuers
Clerical and secretarial occupations	A good standard of general education. Certain occupations will require further additional vocational training to a well defined standard (e.g. typing or shorthand)	
Craft and related occupations	A substantial period of training, often provided by means of a work based training programme	Bricklayers; glaziers; plasterers; woodworking trades
Personal and protective service occupations	A good standard of general education. Certain occupations will require further additional vocational training, often provided by means of a work based training programme	
Sales occupations	A general education and a programme of work based training related to sales procedures. Some occupations require aditional specific technical knowledge but are included in this major group because the primary task involves selling	
Plant and machine operatives	The knowledge and experience necessary to operate vehicles and other mobile and stationary machinery, to operate and monitor industrial plant and equipment, to assemble products from component parts according to strict rules and procedures and subject assembled parts to routine tests. Most occupations in this major group will specify a minimum standard of competence that must be attained for satisfactory performance of the associated tasks and will have an associated period of formal experience related training	

Table 9.4 contd

Major group	General nature of qualifications, training and experience for occupations in the major group	Typical examples in construction
Other occupations	The knowledge and experience necessary to perform mostly simple and routine tasks involving the use of hand-held tools and, in some cases, requiring a degree of physical effort. Most occupations in the major group require no formal educational qualifications but will usually have an associated short period of formal experience related training. All non-managerial agricultural occupations are also included in this major group, primarily because of the difficulty of distinguishing those occupations which require only a limited knowledge of agricultural techniques, animal husbandry, etc. from those which require specific training and experience in these areas. These occupations are defined in a separate minor group	

Source: adapted from HESA *First Destinations Circular* (June 2001)[19].

form of occupation. As the latter move towards becoming professions they are said to be professionalising.

Professions are usually held in high esteem by society and many occupations seek to attain professional status. The UK Government's occupational grouping under the Registrar General places professions in social category I on a six-point ordinal scale. A common feature in the growth of a profession is the demand for a particular service, and there is an increase in the level of formal provision as the service grows and becomes more recognisable in terms of its skills, inputs and outputs. Consequently the limits of the knowledge base become more established. This can be a positive as well as a negative development, turning on inclusivity versus exclusivity or those that are 'in' versus those that are 'out'. A further feature of the growth of a profession is the monopoly, power and influence accrued by it in society and consequently the price that it can charge for its services in the market place.

Professions, exercising a form of control over highly specialised skills in the occupational hierarchy, are also involved in social, societal and market based relationships. The potential for high incomes in the professions stems from their power, prestige and status. In order to be perceived and accepted as a profession an occupation requires public recognition. Professionalism is an occupational ideology of those that are traditionally called 'professions' such as medicine and the law. It is often used by occupations seeking to gain prestige and status.

Professionalism is largely a strategy used by an occupation to seek, gain or retain occupational power by means of the system of 'credentials' offered through its institutional associations. However, there are many types of 'professional associations', from those that are essentially trade associations representing the interests of their members, to learned societies, through to associations concerned with ensuring that individuals are fit to practice. Some professional associations combine several of these functions.

Professional associations usually confer credentials to practice and the most important for professional behaviour is the qualifying association. The qualifying association exercises considerable control over the occupation, tests competence and consolidates and develops the profession's knowledge base. Qualifying associations also provide control over recruitment and training. They are normally involved in the educational process of those seeking to enter the profession. These factors in combination – the exercise of occupational control by qualifying associations and the knowledge base resting on expertise – provide an important power base for certain professions. Historically, society has granted certain professions autonomy to self-regulate and this is normally a profession's best claim to status. Many of the established professions not only self-regulate but also have codes of conduct to regulate professional behaviour.

Over time sociologists, occupational psychologists, organisational analysts and business strategists have analysed professions from different perspectives and a number of models of professions and professional behaviour have emerged. The *trait approach* attempts to identify common attributes separating professions from other occupations. Archetypal traits of 'professions' are seen to be:

- The capability to self-organise through professional associations.
- Members adhering to a code of conduct.
- The requirement for training and education, normally to degree level for prospective entrants.
- Being seen as a complex occupation with a knowledge base that requires the exercise of specialist skills.
- A set of skills that are based on and derived from theoretical knowledge.
- Members holding a sense of vocation and belief in altruistic service.

Proponents of the trait approach argue that occupations can possess more or less all of the above in their progress towards becoming a profession. The trait approach dominated early research on the professions during the 1960s and 1970s.

Political economists also became very interested in professions as a specialist occupational grouping because of their linkages to power and status structures in society that resulted in the potential to accrue high incomes. The *market closure* perspective argues that the occupational strategy of professionalism confers market power on an occupation. This power is derived from the credential awarded to members of the profession by their professional associations, and through this professions are able to exert market closure and control the number of prospective candidates entering. They enforce exclusivity and this enhances the market value of the profession through the forces of supply and demand. Within this perspective three forms of institutionalised occupational control are recognised, namely *professionalism*, where the power balance favours professions compared to their clients; *patronage* and its modern equivalent *corporate patronage*, where the power balance favours clients rather than the professions serving them; and *mediation*, where the institutional power structures of government (or some other powerful mediating agency) intercede in the relationship between the professions and their clients. The first two, professionalism and corporate patronage, are particularly relevant to this debate.

Other researchers have been interested in the skill and knowledge base of professions and the link to and balance of power between practitioner and client. Occupations are alleged to possess a combination of cruciality and mystique. An occupation has *cruciality* when perceived as vitally necessary for the prosperity or survival of 'significant others' that use the services of the occupation, namely its clients. *Mystique* develops from cruciality and accrues when an individual or organisation seeks out a practitioner seen as crucial for solving a problem. Mystique – the 'aura of mystery' surrounding an occupation – is created when the practitioner is perceived as possessing extensive knowledge on the relevant subject in comparison to the lay person. This creates an authority relationship in favour of the practitioner. Cruciality and mystique work in tandem such that the greater the cruciality and mystique of an occupation, the higher is its status as a profession.

Finally, some researchers have focused on the processes operating within professions, arguing that professions are made up of different segments of practitioners. This is the *occupational segmentation* perspective. Proponents of the segmentation perspective argue that professions are coalitions of major segments of practitioners who are organised around a particular specialism. Each segment has its own ideology. Within a particular profession loose amalgamations of practitioner segments can exist and operate under a common institutional umbrella. The medical profession is one example, the surveying profession another.

The professional association, existing as an organisational form in its own right and with its own structures, represents the locus for competing power groups within professions. The professional association is the arena where they come together at an institutional level to present their own view of professionalism to the outside world. Over time the fortunes of powerful coalitions change as individuals come and go and the profession, and its ideology, evolves and changes to reflect adjustments in those power groupings.

The surveying profession is an example in construction of the dynamics of segmentation. The structure of the Royal Institution of Chartered Surveyors, the primary qualifying association for the surveying profession, has changed over time to reflect different specialist interests. Male concluded in 1984 that the surveying profession covers a wide diversity of occupations. Historically, a diversity of professional associations represented segmental interests. One example of segmentary influences impacting institutional structuring has occurred between the RICS and quantity surveying. Within the RICS there had long been an uneasy relationship with quantity surveying which was recognised as virtually a separate profession in its own right[21]. Power struggles within the RICS resulted in two new associations forming to represent the interests of quantity surveyors: the Quantity Surveying Association in 1903 and the Institute of Quantity Surveyors in 1939. Many of the tensions were as a direct result of quantity surveying emerging from the architectural profession and the existence of differing ideological mindsets between those quantity surveyors working in private practice and those working in contracting. The former argued that they operated under an ideology of professionalism and worked directly for clients whereas the latter operated under a system of commercialism. Subsequently the RICS and IQS reunited. More recently, in 2001, the divisional structure of the RICS was superseded by a new structure comprising 16 faculties representing new specialisms within the surveying profession – none of which is termed 'quantity surveying' – including a faculty termed management consultancy. Membership of certain faculties permits members to use alternative chartered surveyor designations.

The following sections explore the implications of the preceding models on the continued development of value management.

Inter-occupational rivalry, professional associations, value management and professionalism

An earlier section has argued that it is difficult as yet to define value management as a separate occupation. It is offered as a service under different practice settings which have been outlined earlier. There is a diverse set of institutional influences on VM in the UK comprising European and UK pressures through the Institute of Value Management, the construction professional institutions and academia. There have been other influences on its deployment in the UK such as an HM Treasury CUP guidance note, a European/British standard for VM, the Latham Construction Industry Board Special Interest Group, the European Union SPRINT programme, endorsements from the Egan Report *Rethinking Construction*, and interactions with practitioner associations and academics in other countries. These institutional frameworks simultaneously expand, clarify, amplify, confuse and add to the rich development of value management in the UK and internationally.

UK construction industry professional institutions that have laid claim to or are expressing an interest in value management are:

(1) The Royal Institution of Chartered Surveyors (RICS) has stipulated that VM is one of its core services and has published numerous research papers on VM. It is interesting to note that it has established a management consultancy faculty which could be the natural home of value management.
(2) The Chartered Institute of Building (CIOB) has published research papers on VM.
(3) The Institution of Civil Engineers (ICE) has produced a Guidance Note on VM.
(4) The Chartered Institution of Building Services Engineers (CIBSE) has produced a Guidance Note on VM.

(5) The Association of Project Managers (APM) has a special interest group for VM.

The Institute of Value Management, encompassing construction as well as other industry sectors, has indicated that management approaches such as business process reengineering, total quality management and other related value orientated methods should also be part of the value manager's armoury. By broadening the methods and toolbox of the value manager into these areas the Institute of Management is also potentially brought within the ambit of those professional associations having an interest in value management. It also recruits members from the construction professions.

The RICS, CIOB, ICE, CIBSE and APM are all qualifying associations. They test competency. They are also actively involved in consolidating and developing the knowledge base associated with their own occupational domains. As qualifying associations the RICS, CIOB, CIBSE and ICE, in particular, all exercise occupational control over entry and the award of credentials at undergraduate degree level. As part of this process they will accredit or recognise undergraduate degree courses in tertiary educational establishments which meet their institutional criteria. Members of these professional associations who practice as value managers have already gone through a certification process for their own professions. In addition, members of the APM with an interest in VM will often have multiple professional qualifications from different institutions due to the broad nature of project management. Members within these qualifying associations with an interest in or who practice value management are consistent with the segmentation perspective of established professions. This poses a problem for the Institute of Value Management in the UK, recognised by the EU Commission as the appropriate body dealing with value management matters. The IVM is not a qualifying association in the same way and with the same stature as the RICS, CIOB, CIBSE and ICE who all confer chartered status on their members. This is explored further below.

Education and training

Education and training of practitioners at university level is closely allied to the professionalisation process. In the UK construction industry VM education is now embedded within the tertiary educational fabric of the country, at both undergraduate and postgraduate levels. It will normally be delivered as a specialist module or modules offered within a wider degree structure. Educational provision in value management at undergraduate level is likely to be at the level of awareness only. However, at postgraduate level it is highly likely to be a combination of both education and training. Universities, through their Statutes and Charters, are able to offer degrees in areas where they have expertise, where there is a potential demand for recruits or where there is a society expectation that certain degree courses provide an important contribution to the vocational or cultural fabric of the country.

Course development and delivery will often be linked to developments in professional practice or research. This is especially the case within the knowledge domains of the traditional professions. In the construction area there is normally a close relationship between the university sector and the relevant professional qualifying associations. In practice a symbiotic relationship exists between academia and the qualifying associations operating to the mutual benefit of both. The professional associations influence courses; equally, there is reciprocity in that academics are often members of committees within professional associations which in turn influence their development. Universities also carry out a research and development role for professions. Thus the link between the theoretical base of a profession and the education, research and training base within the university sector is important for the development of a profession.

There is no doubt that value management in construction has been well established in the tertiary educational sectors of the UK and Australia for some time and in the USA to a much lesser extent. The traditional construction qualifying associations in the UK have been accrediting courses with value management embedded within the curriculum, certainly at undergraduate level, for a number of years. In the same vein many have been sponsoring research studies in value management to support the continued development of the profession; the RICS is a case in point. The results of this research have been incorporated into course development at undergraduate and postgraduate levels.

The Institute of Value Management, under the auspices of the European Governing Board for value management recognises a three-stage process for training practitioners covering nine days of teaching, seminars, case study and practice:

(1) The foundation module: an awareness module only.
(2) Advanced 1 module: providing sufficient material to permit an individual to co-lead a study.
(3) Advanced 2 module: providing sufficient material to permit an individual to lead a study.

There is no formal academic or experience requirement to enter onto the basic module. However, there is a requirement to participate in two value management studies before undertaking the advanced 2 module. Each of the modules identified above must be ratified by the IVM's Certification Board in terms of content and delivery methods. The IVM's Certification Board is an independent body whose role is to implement and control the certification and training policies developed by the IVM and the European Governing Board (EGB). The latter represents all European value associations. These training courses can be offered by independent certified trainers. The IVM website indicates that the qualification of trainer in value management is the recognised qualification at European level for this. The qualification is only accessible to professionals in value management (PVM) with at least two years' experience. The PVM qualification is the recognised qualification at European level for those capable of leading a range of value studies in different environments. In parallel to these training initiatives at institutional association level, the charters of universities provide for the delivery of short continuing professional development courses such as those identified for value management above. This may inevitably bring academics into direct competition with practitioners who deliver similar courses from within their own personal and organisational bases of expertise.

The interactive relationship between the university sector and the institutional qualifying associations means that there are checks and balances in place for the quality of delivery. These occur at two levels: first, within the institution itself through its course vetting procedures; and second, when the qualifying associations visit university departments to accredit courses, normally on a five-year rolling basis. Shorter duration continuing professional development training courses will, however, normally stand outside this accreditation process. Certain qualifying associations in construction have approved centres for continuing professional development in the university sector. At postgraduate level they may well also accredit one-year or two-year Master's degrees.

To conclude, value management falls within the sphere of influence of many traditional qualifying associations in construction. The higher education sector, important for the continued development of a profession, can deliver courses autonomously. From a UK perspective it could be argued that the university sector, with a well established bedrock of teaching in value management at undergraduate and postgraduate levels could proceed alone in educating and training VM practitioners. Normal practice in higher education would suggest that any number of qualifying associations could accredit courses in value management. This could also include the IVM Certification Board or the SAVE International equivalent under their global outreach programme.

Ethics, codes of conduct and professional negligence

One of the other defining characteristics that differentiate professions from other occupational groups is adherence to a code of conduct. This sets the ethical boundaries for professional behaviour.

Professional behaviour is usually characterised by making decisions that are in the best interests of the client and not the practitioner. Decision making involves evaluating alternative courses of action and requires a set of basic values to be used in comparing the merits and outcomes of each. Armstrong, Dixon and Robinson argue that ethical decision making is about the quality of the decision, taking account of justice, equity, the consequences for those affected by the decision and with the personal and collective responsibilities that lie behind any moral or contractual obligation that is entered into. Ethical decisions are concerned with what is 'good' and 'right' and dealing with conflicts between rival 'goods' and 'ills'[22]. Armstrong *et al.* add further that ethical problems do not have easily defined solutions but are usually surrounded by ambiguities, complexity and ill-defined boundaries. In this sense ethical decision making requires high-level cognitive and judgemental skills, normally seen as the domain of the 'professional practitioner'. They argue that professional codes are important for ethical, reflective development and are the mechanism for developing the integrity of the collective of practitioners. In this context one of the important policy issues to have emerged from the benchmarking study in terms of value management is the relationship between certification of value managers, legislating for the use of VM, the development of standards and institutional associations.

When looking at those institutional associations whose sole remit is value management, the IVM indicates that it aspires to establish its role in VM as[23]:

- As a brand enhancer to raise awareness and promote the use of value management in all sectors of the UK economy.
- As a learned society facilitated by the Executive Committee and Special Interest Group to promote research and increase VM knowledge.
- As a regulator operated by the Certification Board to establish a nationally recognised set of standards and skills within the EU framework.

The Code of Ethics of the IVM indicates that a professional VM practitioner is expected to act in accordance with the following precepts:

- Clearly identify the needs and expectations of customers and other stakeholders and aim to fulfil or exceed them.
- Notify the customer if there is a mismatch between the needs and expectations and the means to achieve them.
- Keep abreast of their subject and contribute to its development.
- Help others to reach professional fulfilment as well as promote wider public understanding of value management and the success it can achieve for the customer.

The SAVE International website indicates that its mission is to lead and expand the value profession by[24]:

- Fostering education.
- Communicating VM news and activities.
- Promoting the ideals of professional practice.
- Broadening the application of VM use.
- Recognising significant VM contributions.

While it does not publish a code of ethics per se on the website it acknowledges the following as core values and beliefs which set its boundaries in the pursuit of its vision:

- Foster an environment for personal and professional growth.
- Embrace honesty and integrity.

- Celebrate the accomplishments of members.
- Advance the profession worldwide.
- Concentrate on strengthening the knowledge of members.

Both have codes of ethics (termed core values and beliefs by SAVE International) and both have certification boards. The language used by each association is interesting. SAVE International argues its role is to *lead* and *expand* the value profession. It lays claim to professional status with the use of the word 'profession'. Its members believe they are there already, perhaps because the US Government requires that certified value specialists, the title conferred by SAVE International on certified practitioners, must be used on government work. The IVM *aspires* to establish itself as a brand enhancer, presumably meaning that it provides an additional skill set beyond those already acquired in other occupations and professions; as a learned society to promote research and increase VM knowledge; and as a regulator of practice within a wider European framework. In terms of the certification of practitioners of value management the important issue is that good ones are seen to be in short supply. They may be self-selecting. Certification does provide the client or end user of the service with peace of mind that they are using a qualified individual.

Professions are normally granted by society, expressed through the legal or governmental frameworks of a country, the autonomy for their qualifying associations to certify for practice using a particular designation, e.g. chartered civil engineer, chartered surveyor or architect (a legally protected title). As indicated earlier, these associations of practitioners define what is the knowledge base and what are the accepted tools and techniques for use in practice. A number of countries, however, have standards for value management including Austria, Australia and New Zealand, France, Germany, Hungary, India, Poland and the USA (through SAVE International). EU countries with value associations have adopted the European Standard.

But why have a standard? There is no British Standard, for example, for architecture or for quantity surveying but architects and chartered surveyors have designations guaranteeing the competence of holders of that title. It can be deduced that a standard is necessary for three reasons. First, the service of value management is sufficiently vague in the minds of those who commission it to need clarification. Second, the institutional framework of practitioner associations is insufficiently well developed for society, through a legal or governmental framework, to entrust them with the autonomous power to certify and self-regulate. Third, practitioners see it as important for establishing the boundaries around a particular job activity but do not have the collective organisation with power in the marketplace to enforce certification through a qualifying association.

However, the development of standards is a double-edged sword. Practice can be judged for adequacy against a benchmark and this may set the boundaries for claims of negligence. Standards are also useful in defining, reinforcing or clarifying terminology for wider acceptance but VM is widely recognised as a change orientated process. The use of standards may proceduralise practice, hinder innovation and deter experimentation, a vital ingredient of methodological development.

A study conducted by the authors[25] determined views on the limits of a value manager's culpability for negligence in conducting VM studies. One view is that the value manager's responsibility is to apply the value methodology while acting as a process manager and facilitator of workshops. There are two possible contexts for a value study:

(1) Working with a team of record.
(2) Working with an independent team.

In either case the outcomes of a study are the result of team working and the value manager reports on these. This signals that the value manager is also responsible for the accuracy of the report.

A differing perspective is that if a value manager is qualified in a different professional discipline, for example as a chartered civil engineer then that value manager may not only be found negligent in the application of the VM methodology but also potentially in that of his/her base discipline if an error should come to light within the civil engineering area which was known about by the value manager and not reported. A further point that has recently been raised in the area of negligence is the increasingly close affinity between value and risk studies. The point at issue is that since risks are insurable then potentially, if a value study fails to identify, misclassifies or fails effectively to mitigate a risk because of an inappropriate application of the value methodology, the value manager picks up this liability also.

9.5 Summary and conclusions

The discussion in this chapter has been far reaching, touching on occupations and the social structure, the nature of professions and the location of VM in such a framework. Earlier sections have argued that VM is a method that comprises a process – the structured value study plan – and that function analysis is its founding and internationally well-recognised technique. The academic base of VM as a discipline is in its infancy and it draws on many tools and techniques from other disciplines. It has become embedded in the higher education sector of some countries to a greater extent than in others. Internationally it has been dominated by practitioner developments and does not have a unique theoretical underpinning. It has, however, been founded on a unique technique. The structured process for using that technique is a good example for problem solving in general. A foundation on a technique(s) and a process alone are insufficient for a claim to professional status. To attain professional status the important issues are first, that a distinctive occupational role has been established; and second, that practice is guided by a unique, dedicated body of theory that is taught at university level. Currently VM cannot claim this.

Empirical evidence indicates that VM at present should not be considered any more than a method; it is not recognised within Government job structures defining occupations, for example, in either the UK and or the US. However, it is argued here that the methodology requires a skill set that necessitates high level cognitive functioning at practitioner level; not just facilitating a workshop, but planning for, conducting and ensuring study outcomes are implemented to the benefit of the commissioning client and stakeholders. To do this requires the integration of knowledge, mastery of the process, tools and techniques, flexibly to meet a diversity of value problems (strategic, organisational and operational), advocacy to a wide audience and ethical decision making. Value management has been endorsed by governments in Australia, the UK and the USA, to name but a few, as good practice.

Defining the limits of an occupation and profession is not easy. The term profession is surrounded in confusion, it is often overused and is probably best kept as a term for a particular occupational phenomenon where the business and social worlds overlap. There are a number of different views on the nature of professions. The underlying themes are associated with power, status, qualifications, education, training, and market inclusion and exclusion through the credentials offered by institutional qualifying associations. The esteem in which the credentials, professional associations and practitioners are held is an important factor in terms of the recognition to a particular profession by society.

A number of conclusions can be reached about practitioner skills. First, the evidence strongly supports the view that good practitioners are in short supply. Those that are seen to be good in the marketplace have an opportunity to establish an authority relationship between the practitioner and client based on cruciality; that is, they are seen to be delivering a service which the client sees as critical. Second, the use of high level cognitive skills varies along the continuum of practitioners, from those that facilitate workshops only through to those that operate on the basis of a comprehensive service covering an extensive orientation and diagnostic phase, a flexible and adaptive workshop phase and a focused reporting phase.

The economics of the market place are driving VM to a shorter study duration, and hence a less comprehensive service. Often this can be as little as facilitating a workshop. In the marketplace this is verging on a 'commodity' service. Hence, it is argued here that facilitation styles 1 and 2 operate at the level of 'artisan', requiring skills that can be learnt through experience; facilitation style 3 operates at the boundary between 'artisan' and 'professional'; and facilitation style 4 operates at 'professional' level. In the latter case there is the potential for good value managers to create a mystique around the value management process since not everyone is capable of doing it and it requires high level cognitive skills to work effectively throughout the process phases of orientation and diagnosis, workshop and implementation.

The high degree of mystique and cruciality potentially surrounding the value management process could enable a sector of good VM practitioners to establish a sound professional base for value management backed up by a robust certification/qualification system. To be effective and recognised in the marketplace the qualification must be seen as a mark or standard of knowledge, skill and ability to practice. However, this would also require an effective policing system to define what is good VM practice and what is not and hence differentiate the good from the not-so-good practitioner. This is important for the continuing reputation of the service, for certified practitioners and for regular or irregular commissioning clients of the service. It also provides the basis for market closure and identifying those that are 'in' and those that are 'out'.

At present in the UK, chartered institutes with an interest in VM are already well embedded with many of the traits of a profession and the associated trappings of professionalism, housing VM among their specialist services within a portfolio of services. However, to the best of the authors' knowledge none of these have specific training and certification programmes dedicated to VM, although this was recommended to the RICS in 1986[26]; they certify within their core disciplines, with components focusing on value management. The certification body of the IVM administers the European Certification Scheme in the UK on behalf of the European Governing Board. Currently the secretariat is provided through the IVM and this does have the advantage of representing value management in all industrial sectors.

To add to the richness of the territorial debate over value management, SAVE International launched a global outreach programme at its 1997 Conference with the specific aim of widening its influence and membership internationally. This adds a further dimension to the institutional framework within which VM is developing. As the original founding association for a value orientated approach, coupled with the fact that there are a number of SAVE franchised associations internationally, they may become an additional important player on the UK scene, especially if they accredit a range of university courses with specialisms in VM.

The foregoing suggests no clear picture as yet of an institutional framework for VM development in the UK. What is clear, however, is a potential for territorial 'turf warfare' through inter-professional competition counterbalanced by the fact that the certification for VM is currently managed on behalf of the EU by the Certification Board of the IVM.

Clearly the evidence supports the fact that VM practitioner segments are present within the existing construction professions, and these are evidenced by their professional qualifying associations. Equally, within the IVM practice segments have already begun to emerge with the establishment of four special interest groups. Taking the long term view, it is unclear as yet which of the many institutes may form the natural home of value management or whether any one Institute can claim an exclusive territorial right over it. However, what is certain is that institutional interest is moving forward, especially among some of those that have power and influence in the UK construction industry because of their historical longevity. Awarding credentials in the form of professional qualifications, professional responsibility and ethical behaviour is the key to professionalism irrespective of the institutional home. A number of the construction associations with an interest in VM, and mentioned above, already have Codes of Conduct in place and a longstanding view of professionalism within their own expert domains. SAVE International has established a series of core values and the IVM has adopted a Code of Ethics.

An argument can be set out that value management could be considered a profession in a country where the institute responsible for it:

- Becomes recognised by those practising value management as their only advocate and by those procuring value management services as their point of reference; as the accrediting body for courses; and as a learned society for recognised qualifications, i.e. the qualifying association.
- Is run quasi-democratically by the membership.
- Has a code of ethics.
- Has a clearly defined knowledge base and set of tools and techniques that is in a state of continuous ongoing development and is capable of differentiating it from other specialist disciplines.

In this situation the value manager may offer no advice, provide no design nor solve a complex problem. The value management facilitator is a process manager using various tools and techniques including function analysis. The value manager provides a service and is liable only for that service.

However, if VM moves along a continuum such that higher level cognitive skills are essential to design, run and implement a study on a regular basis and value managers are held accountable for a total service package beyond just workshop activity, then it can become a profession in its own right. This will require the full infrastructure of a qualifying association to be in place, graduate or postgraduate entry, and clear boundaries around the knowledge and practice base. The corollary is that if VM attains professional status then a standard is not required since certification is achieved through an autonomous, self-regulating, independent qualifying association. A value manager operating to a particular national standard, however, operates within an occupational framework where service has been ill-defined and recipients need some protection. Table 9.5 uses the trait approach to analyse value management as a service.

No one institute has the power of market closure to control the recruitment and practice of VM. If the VM community has ambitions to develop it beyond a management methodology and follow the path taken by traditional professions then universities remain the best placed institutions to continue its further development, given the existence of standards defining the content of practice and that the European Certification System provides a training schema for academics. Within the market closure perspective VM operates certainly under a system of corporate patronage since particular corporate clients select individual practitioners.

Table 9.5 A trait analysis of VM.

Trait	Comments
Self-organisation through professional associations	In construction there are at least five associations with an expressed interest in VM. Of these some have well-established strategies and ideologies of professionalism in place. This is consistent with a segmentation view of those professions. The IVM is seen by the EU as the appropriate body for dealing with VM related matters. It does not have a well-established system and ideology of professionalism in place. In comparison to the chartered institutes, the IVM has closer similarities to the APM which also recruits across industrial sectors. Equally a number of the chartered institutes have long-standing interests in project management. VM cannot claim as yet to have a clear institutional framework for self-governance in place
Adherence to a code of conduct	The chartered institutes have established codes of conduct. The IVM also has a code of ethics. The test comes if the code of ethics is used to disbar a practitioner from practising
Education and training normally to degree level	The chartered institutes have long-established links as qualifying associations with the higher educational sector. They also look to this sector, in addition to professional practice, to deliver training courses on a regular basis. VM has many affinities with PM, in that it is likely to be a graduate post professional activity, given the requirement for experience. This suggests that VM requires a postgraduate delivery process. In the UK there are now many master's degrees in project management that have been established and running for, in some cases, decades. The same argument could be applied to VM; the problem would become one of a specialist series of modules off other courses
Complex occupation with a knowledge base requiring specialist skills	The earlier analysis clearly indicates that VM does require the use of high level cognitive skills, often to solve complex value related problems, provided it is not just seen as facilitating workshops. While it does require specialist skills they are not, with the exception of function analysis, unique to VM and other occupations/professions can lay equal claim to the knowledge base once function analysis has been mastered
Skills based on theoretical knowledge	This is clearly the case but the theoretical knowledge base is not unique to VM. Again, other occupations/disciplines could lay claim to the knowledge base of VM
A sense of vocation and belief in altruistic service	Most professions lay claim to altruistic service and this is probably best expressed as doing the best for the client in a business context. Empirical evidence from a previous study identified that a sense of vocation increases with age. It is probably a function of the opportunity cost of time and effort already sunk into a particular vocation compared to that of taking up another or having transferable skills. VM is probably one occupation that relies heavily on transferable skills

An earlier section has also argued that the use of high level cognitive skills throughout a value study (orientation through to implementation) in some circumstances has the capability to enhance cruciality and mystique. The benchmarking study identified empirically that one major problem standing in its way is that in certain organisations in the UK, Australia and the US VM has become a box-ticking exercise. This suggests it is performing merely a routine function within those organisations rather than being used to bring innovative change from a multi-disciplinary team.

The conclusion from this analysis is that VM, as a methodology, is a structured process which within limits is consistent internationally; it relies on function analysis as its principle founding and differentiating technique and has all the signs and trappings of professionalising. Much of the practice currently falling within the ambit of the term value management is at the level of a craft/ artisan capability[27,28]. Professional development relies on a stricter control of recruitment and certification, setting the boundaries and codification of the knowledge base, and moving education and training into the university sector with VM institutes becoming qualifying associations.

9.6 References

1. Male, S. P. (1990) Professional authority, power and emerging forms of 'profession' in quantity surveying. *Construction Management and Economics*, **8**, No 2, pp. 191–204; Male, S. P. & Kelly, J. R. (1990) The economic management of construction projects: an evolving methodology. *Habitat International*, **14**, No 2/3, pp. 73–81; Male, S. P. & Kelly, J. R. (1989) The organisational responses of two public sector client bodies in Canada and the implementation process of value management: lessons for the UK construction industry. *Construction Management and Economics*, **7**, No 3, pp. 203–216.
2. Kelly, J. R. & Male, S. P. (1993) *Value Management in Design and Construction: The Economic Management of Projects*. E. & F. N. Spon, London.
3. Male, S., Kelly, J., Fernie S., Gronqvist M., Bowles G. (1998) *The Value Management Benchmark: A Good Practice Framework for Clients and Practitioners*. Published Report for the EPSRC IMI Contract. Thomas Telford, London.
4. Male, S. P. & Kelly, J. R. (1999) The professional standing of value management: a global study of legislation, standards, certification and institutions. *Proceedings of SAVE International Conference, San Antonio*, pp. 158–166.
5. BS EN 12973:2000 *Value Management*. British Standards Institution, London.
6, 7, 8, *Value for Europe: Training and Certification System Manual* (1997) IVM, London.
9, 10. SAVE International (1988) *Value Methodology Standard*. Revised October 1998. Society of American Value Engineers, Dayton, Ohio, USA.
11. *Value management guideline*, January 2001. DPWS Report Number

01054, Total Asset Management Manual (TAM 2000), www.gamc.nsw.gov.au/TAM2000/
12. Male, S., Kelly, J., Fernie, S., Gronqvist, M. & Bowles G. (1998) *The Value Management Benchmark: Research Results of an International Benchmarking Study.* Published Report for the EPSRC IMI Contract. Thomas Telford, London.
13. Egan, Sir John (1998) *The Egan Report: Rethinking Construction.* Department of the Environment, Transport and the Regions, London, Publications Sale Centre, Rotherham, England.
14. Building Research Establishment Research Project.
15. Male, S. P. & Kelly, J. R. (1999) The professional standing of value management: a global study of legislation, standards, certification and institutions. *Proceedings of SAVE International Conference, San Antonio*, pp. 158–166.
16. Miles, L. D. (1989) *Techniques of Value Analysis and Engineering*, 3rd edn. Lawrence D. Miles Value Foundation, Washington D. C.
17. Hall, R. H. (1969) *Occupations and the Social Structure.* Prentice Hall, Harlow, Essex.
18. Thomas, R. http://qb.soc.surrey.ac.uk/resources/classification/socintro.htm 8 February 2003.
19. HESA (2001) *First Destinations Circular, June 2001.* Higher Education Statistics Agency, Cheltenham.
20. Male, S. P. (1984) *A Critical investigation of professionalism in quantity surveying.* PhD thesis, Heriot-Watt University.
21. Monopolies and Mergers Commission. Surveyors Services (1977). *A Report on the Supply of the Surveyors Services with Reference to Scale Fees.* HMSO, London.
22. Armstrong, J., Dixon, R. & Robinson, S. (1999) *The Decision Makers: Ethics for Engineers.* Thomas Telford, London.
23. http://www.ivm.org.uk/aboutivm_mission.htm, 16 February 2003.
24. http://www.value-eng.org/about.php, 16 February 2003.
25, 26. Kelly, J. R. & Male, S. P. (1986) *A Study of Value Management and Quantity Surveying Practice.* Surveyors Publications, London.
27. Male, S. P. & Kelly, J. R. (1999) The professional standing of value management: a global study of legislation, standards, certification and institutions. *Proceedings of SAVE International Conference, San Antonio*, pp. 158–166.
28. Woodhead, R., & McCuish, J. (2003) *Achieving Results.* Thomas Telford, London.

10 The Future of Value Management

10.1 The development of value management

The early years of value management have been dominated by US practice. A watershed occurred in the mid 1980s with the international use of the method in construction. While there has been interest from some countries in moving VM forward by franchising the US methodology it has also been taken and melded into a diverse range of international construction markets and cultures. At this juncture contextualisation has occurred within these markets and has forced further developments in value management thinking and practice. It has also resulted in a diversity of definitions and procedures within official standards. Value management is still developing, but there are signs that the development is slowing – some might say stagnating – being driven principally by tick-box approaches by clients and the economic drive towards shorter studies. This conclusion takes value management into a series of possible future scenarios for its further development. These are elucidated further in this chapter.

The analysis undertaken in this book also confirms the view that value management is currently a management method offered by a range of 'consultancies' in different practice settings. Function analysis is the technique that sets value management apart from other management methods. Function analysis drives the determination of value through an explicit exposition of why something exists, what it does and what is must perform. It then seeks to relate options/alternatives to solve function in relation to comparative cost. Value comes to the fore when choosing the appropriate alternative against a set of predetermined criteria: a value system.

Currently in the UK many professional institutions lay claim to VM as a service or, at a minimum, have a special interest in it. This typifies a segmentated approach to professional practice where there is a lack of clarity over institutional and practice boundaries and where one institution alone is unable to ensure market closure for recruitment through certification. It is not necessary to have a certificate to practice value management in the UK although in certain countries certification is a requirement for undertaking the service on public sector work. Exploration of a typology of practice setting has identified that value management is a service that does not necessarily require a significant supporting infrastructure but does require an effective network in place to handle different workloads and also to obtain work.

It is argued here that the development of value management beyond just a methodology that includes workshop facilitation should embrace a more comprehensive package of skills. The future lies in a more holistic and inclusive service which incorporates greater involvement in the orientation and diagnostic planning of projects, using the workshop phase to reconcile and coalesce different value systems and cultures and the implementation phase to ensure that options or alternatives to enhance value are actually implemented. This becomes an advice-laden professional service in which the value manager must accept liability for his/her professional advice. Offering a professional service and taking on board a wider liability for that service raises a tranche of issues for certification and practice Standards, such as BS EN 12973:2000 and the SAVE International Standard.

10.2 An enhanced VM process

A useful characterisation of practice is included in BS EN 12973:2000 which identifies three generic study styles:

(1) A strategic study.
(2) An organisational study.
(3) An operational study (project focused).

In addition, the authors have defined a framework for undertaking a value management study. This framework comprises three interlocking phases:

- The orientation and diagnostic phase which includes appropriate selection of study style and tools and techniques to elicit and restructure information for decision taking.
- The workshop phase, for resolving competing value systems, aligning or re-aligning the value chain, keeping the value thread intact and generating ownership of function and solutions.
- The implementation phase where it is ensured that options to enhance are implemented.

The above must be tailored to take account of the different styles that may be available for a particular study. Table 10.1 presents an enhanced VM process taking account of observations made from the chapters in this book and a review of the standards identified in Chapter 9.

The emphasis of each of the major phases and the stages within them varies depending on whether it is a strategic, an organisational or an operational study. Generically, however, the enhanced process identified above provides a comprehensive progression through and structure to a value study.

The next section outlines a structured framework for using value management to deliver best value in a public service context.

Table 10.1 An enhanced value study process.

Orientation and diagnostic phase			
Major phases	Stages	Activities	Comments
Orientation and diagnostic phase	Identify value context with project sponsor and stakeholders	Hold briefing meeting with commissioning project sponsor	The orientation and diagnostic phase is concerned with understanding the strategic context of the value study, its scope, timing, schedule and constraints
		Define objectives of value study in relation to commissioning organisation's VM policy and programme	
		Determine what is at stake for the organisation from the VM study	It is important during the early stages of this phase to gain stakeholder commitment and often this may require presentations or confidential interviews
		Identify and gain commitment from stakeholders	
		Agree implementation programme with project sponsor	
	Define VM study scope	Determine study style	The exact nature of the study style will be determined and needs to be agreed with the project sponsor. This will determine the deliverables and the performance criteria for a successful study
		Agree scope and objectives of study with decision maker/project sponsor	
		Agree constraints for VM study – real and apparent	
		Determine VM study evaluation/performance factors	Equally, there may be a series of anticipated studies, for example, at project programme level, and it is important to scope how these will all fit together
		Determine the framework for the VM study	
		Determine the time scale for the VM study	
		Identify if further VM studies may be required	
	Gather comprehensive data for the VM study	Interview VM study participants if using 'team of record'	It is important that the value manager gathers all relevant information. This may include revisiting stakeholders and interviewing confidentially other potential participants if a team of record is to be used. If an independent VM study team is to be assembled it is important that skills are tailored to the problem at hand
		Collect user/customer attitudes	
		Build data models	
	Identify and select team; identify team resources and competencies	Finalise VM team composition and agree with project sponsor	
	Study logistics	Brief participants with a meeting or with prepared documentation	
		Arrange venue	

Table 10.1 *continues*

Table 10.1 *contd.*

Orientation and diagnostic phase			
Major phases	Stages	Activities	Comments
		Distribute and consolidate background material and issue briefing pack, if required	Briefing the VM study team is also important during this phase. This could be undertaken during confidential interviews, through a presentation or through the issue of a study pack. During this phase the value manager will be developing views on what the value problem might be or if there are competing value problems to be resolved during the workshop phase. Equally the value manager should be weighing up the extent to which participants are in agreement or far apart on the value problem(s). An agenda for the workshop phase will be formulated in conjunction with the project sponsor and issued to all participants
		Brief presenters for information stage of workshop phase	
		Commence initial function definition	
	Develop workshop agenda and process	Agree workshop phase agenda with project sponsor	
		Scope team processes and procedures at each stage of workshop phase. Identify working groups, competencies and scope tasks	
		Identify tools and techniques for use during workshop phase	
		Re-confirm VM study objectives and modify scope if required	
Workshop phase			
Major phases	Stages	Activities	Comments
Workshop phase	Information sharing	Confirm value study objectives with VM team	During this part of the workshop phase, the agenda and participants to the workshop will be inroduced. Presentations may be used to commence the issues analysis phase. It is important that the client value system is made explicit at this point and that all participants are fully aware of its implications
		Presentations: scheme and project overview	
		Conduct issues analysis and theme	
		Prioritise issues; identify and explore critical success factors	
		Determine and agree client value system with client representatives	

Major phases	Stages	Activities	Comments
		Workshop phase	
	Back-to-basics: function analysis	Identify and classify functions	This is an important part of the workshop phase; it is central to value management and cannot be rushed. The experience of the authors is that it should not be omitted. Function analysis diagrams will be constructed and other forms of analysis may also be used at this stage, depending on the study style. At the end of this section the team should be able to understand fully *why* something is being undertaken, be it an investment project, a process or spatial configurations. Normally areas for improvement will be identified on completion of the function analysis section
		Build function models	
		Cost functions	
		Establish function worth	
		Select functions for study	
		Conduct other forms of analysis if required: • Process analysis • Spatial analysis	
		Identify areas for improvement	
	Create solutions and generate innovation	Gather existing ideas	Creating new ideas or options is almost always undertaken by brainstorming
		Create new ideas and options	
	Evaluate possible solutions	Rank and rate alternative ideas and solutions for development in workshop	There are a variety of sorting techniques available but the authors have found the method described here for evaluating a quantity of ideas efficient and effective
		Evaluate solutions in terms of: • Client acceptability • Functional suitability • Economical feasibility • Technical feasibility	It is also at this point that working groups are likely to be formed as part of the workshop phase, if they have not been operating in an earlier section
		Identify work programme for developing ideas/ solutions outside of workshop	

Table 10.1 *continues*

Table 10.1 contd.

Major phases	Workshop phase		
	Stages	Activities	Comments
	Present and validate proposals	Work group presentations during final plenary	Working groups can present to each other in a plenary session to cross-validate ideas/ options for final agreement. Often a presentation to executives can be a good focusing mechanism at the end of a workshop to provide further broader inputs into team thinking
		Presentation to senior management, if attending	
	Action planning for implementation	Develop implementation programme: • List activities to be carried out • Identify time frames • Appoint action plan coordinator • Identify action point nominees/champions • Identify any requirements for further VM studies	The development of an action plan commences the process of implementation. VM has been criticised for this major phase of a study and it is essential that action planning is completed before participants leave. Follow-up meetings can also be targeted, especially for presentation of the workshop report to the project sponsor/senior executives
		Identify follow-up meeting/workshop one month after workshop phase complete	
		Finalise proposals, if required	
	Prepare report	Prepare and issue draft report	This aspect can be helped if a recorder is employed to note the workshop proceedings
	Present report and agree final implementation programme	Present draft and oral report to project sponsor	
		Inform VM team of outcomes; dismiss or place on standby	
		Prepare and issue final report	

Implementation phase			
Major phases	Stages	Activities	Comments
Implementation	Disseminate report	Disseminate report to: • Client project sponsor • VM team • Other experts involved in VM study	Experience suggests that the post workshop phase needs further refinement to ensure that implementation of ideas/solutions continues. This could also include further, more detailed working up of options once the workshop has been completed The authors have found that an implementation workshop is very useful to continue the progress of implementation, or at a minimum a meeting with the project sponsor/ senior executives to finalise the outcomes of a study. Often this is integrated with the presentation of the draft report
	Support implementation	Monitor status of actions; follow-up implementation and assist to correct deviations	
		Obtain commitment to implementation programme	
		Hold implementation review workshop two to four weeks after workshop phase	
	Continuous improvement	Collate information on implementation targets and VM study performance	
		Review strengths and weaknesses of VM study with project sponsor	
		Adjust VM study process and procedure	

10.3 Value managing quality to deliver best value

This example outlines a logical and structured framework for the application of value management within the best value programme in UK public services. It describes an integrated quality and value management process of best value that stresses value for money in local authority public services.

'Best value' is the name given to a government policy initiative to improve the provision of local services. The legislative framework for best value is in place in England and Wales and was enacted in Scotland during 2003. The focus of best value is to deliver services to clear standards (covering both cost and quality) by the most economic, efficient and effective means available. Performance reviews of all services are to be completed by 31 March 2005. The review will challenge why and how a service is provided, undertake comparisons across a range of indicators, consult local taxpayers, users and businesses, and embrace fair competition. The information presented here is from a research project that has evolved a three-part approach to the measurement of best value, termed the three wheels. The three wheels are described as follows:

264 The Future of Value Management

- *Wheel 1* considers the assessment of local authority policy and services through the quality schemes currently in force. These include the Balanced Scorecard, Investors in People, Charter Mark, EFQM, ISO 9001, and Six Sigma. The first wheel is applicable to the continuous monitoring of the core business of the local authority, service administration and delivery. In addition, external factors will impact the core business. These include a strategic change delivered by voters, central government legislation, Scottish legislation, rising expectations, competition, cultural change and funding. The continuous monitoring of core business may lead to the examination of particular services by an in-house panel of experts, the exploration and analysis panel (EXAP).
- *Wheel 2* anticipates that the EXAP will proceed with a structured examination of the service which is capable of audit. The most appropriate techniques at this stage are comparison with benchmarking and best practice data, goal and systems effectiveness models, and qualitative and observational studies. The EXAP objective is to determine whether the service is at an acceptable standard or could be at an acceptable standard with an obvious improvement, whether a project is required to determine the form of the service improvement or indeed whether a new service is required.
- *Wheel 3* is the project. In the event that the EXAP raises a service improvement project it will be developed through a process that is value managed. The process will be a structured process capable of audit and will include:
    - The discovery of the service provider's value system by reference to the total quality management statement.
    - The discovery of the customer's value system by reference to the product or service.
    - The function of the service.
    - Innovative strategies that may be used to provide the service in the most economic effective and efficient means available.
    - The preparation necessary to allow the revised or new service to be absorbed into the core business of the local authority.

The three wheel approach to the measurement of best value distinguishes between the review of core business and the best value management of a project. Wheels 1 and 2 have been derived from an analysis of the best value literature. Wheel 3 results from an action research study conducted with a Scottish local authority. An example (Fig. 10.1) has been constructed to illustrate the process, generated partly from the interpretation of literature and partly from action research.

10.4 Value management futures

Having identified the nature and scope of an enhanced value management service that encompasses the developmental requirements for moving from a

The Future of Value Management 265

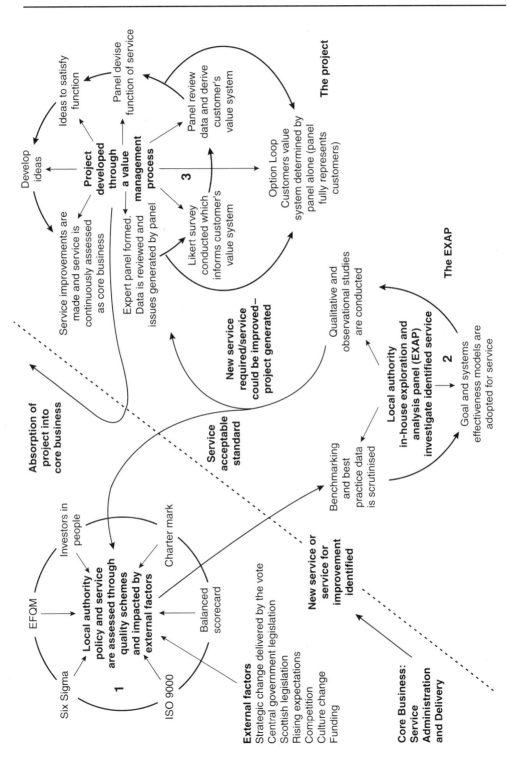

Fig. 10.1 The three wheels of best value: a service focused approach to the measurement of best value.

craft/artisan based activity to one that is capable of becoming a fully professional management consultancy service, this section reviews possible scenarios for continued value management development into the future. Current trends are extrapolated to produce three scenarios; one that represents the status quo, one that represents the most likely case and one that represents a more radical view of the future.

It is difficult to discuss the future of value management in the UK without setting some conditions for the future of the property and construction markets. The authors' views on the development of value management are therefore set against different scenarios of the future within which possible developments in client attitude towards property, procurement strategies, the shape of the construction industry and developments in consultancy are presented. Against these three scenarios possible futures for value management are proposed. Of course the futures under each sub heading are not mutually exclusive but to propose a future for value management for all permutations would be beyond the scope of this chapter. The assumption of the authors is that the truth will lie somewhere between the three.

Scenario 1: the status quo

Construction client attitude towards property

Construction clients generally regard property as the necessary fixed infrastructure for doing business or any social activity. The private sector generally does not see buildings as contributing to administrative, manufacturing or retail productivity and accounts for its cost as an overhead. Manufacturing, in particular, seems happy to occupy old buildings and applies extreme innovation to fitting modern machinery within an outdated physical asset infrastructure. Public sector clients see property as a major part of public service usually designated by its function, e.g. a library, a school, a council building. The concept of the Public Private Partnership (PPP) initiative, where it is not necessary to own the built asset to provide the service, is not being enthusiastically embraced.

Procurement strategies

A large number of clients, particularly in the public sector, rely on traditional procurement methods and 50% of all projects are tendered on full bills of quantities and use traditional forms of contract. Design and build is seen as a means of shifting the risk to the contractor and guaranteeing a fixed price. Framework agreements and public sector procurement systems such as PFI, Defence Estates Prime Contracting and National Health Service Procure 21 are still in their infancy. However, the 2002 report *Accelerating Change*[1] is driving the industry towards 20% of projects by value being delivered by collaborative procurement methods by the end of 2004 and 50% by value by the end of 2007.

The structure of the contracting industry

Cost dominates with only token regard for the value culture of the client. Supply chain management is still in its formative stages. An adversarial climate remains within the industry with lip service being paid to collaborative working. Initiatives tend to come and go at industry level with the majority of initiatives been put forward with missionary fervour on the back of successful projects by major clients. Change management is limited on most construction projects and change is very much accepted right to the end of the project.

The industry is plagued by a poor image at local builder level with little public respect for trade skills. Quoted construction companies are seen as risky investments and generally have low share values when compared to their earnings capacity. A number of construction companies have restructured to become service companies in order to attract a higher share value. Quoted construction companies tend to be management only organisations relying on sub-contractors for construction labour and also expecting them to take responsibility for training.

There is a perception that the industry remains unattractive to new recruits. Evidence exists that school leavers are attracted to the industry but fail to find bona fide training places due to the cost of training for sub-contractors combined with a requirement to bid low to get work from the larger construction companies. Most sub-contractors are lean organisations.

Construction consultancy

The construction profession is huge. For example, approximately one in every 300 of the UK working population is a chartered surveyor. Construction consultancy, however, tends to be split between a few large firms employing several hundred and many private practices employing fewer than 20. The former offer a portfolio of services whereas the latter focus on a core service delivery and attempt to diversify into other markets to secure work. Architectural and engineering practices provide a full design service from concepts to working drawings. Project management has seen significant growth in the past 20 years and is now regarded as normal on most projects of any size. Standardised procedures such as the Standard Method of Measurement and various codes, for example, Co-ordinated Project Information, NJCC codes for tendering, are disappearing but are still used for the 50% of work that is still procured traditionally.

Future of value management under scenario 1

The future development of value management under scenario 1 is limited. It will continue to be used reactively in construction; its full benefits are not understood and it is in danger of being labelled a further cost reduction service rather than a value based service where cost is only one parameter. It will not generally be used proactively nor recognised as a better way of doing business. It is unlikely to be used extensively within the public sector and in this arena will only be accorded 'recommended good practice' status. Value management practice will continue along the lines set out in Chapter 9 with a significant amount of turf warfare between institutional associations as to its appropriate 'home'. However, it is likely to remain a low overhead service provided by sole traders in niche markets.

Scenario 2: most likely

The most likely scenario is developed from evidence of advances that are occurring but are currently emergent or at a small scale. Scenario 2 is likely but may not be regarded as normal before 2015.

Construction client attitude towards property

Some areas will develop to see construction as a commodity, a factor that already exists in parts of the housing sector. The private housing sector is currently atypical in terms of its marketing of a house as a product for sale in a very similar manner to a manufactured product. Within the housing sector considerably more land will become available for construction through the clearing of brownfield sites at public expense and also the making available of more land currently designated as green belt. Major markets will develop in integrated complexes with retail and leisure facilities that contain condominium apartments aimed at young singles and the active retired. There will be significant investments in sheltered housing and care facilities for the elderly within the housing association environment.

The likely future sees buildings as necessary infrastructure that contribute to the efficiency and productivity of an organisation. The financial, administrative and retail sectors will require buildings to be configured to exactly match their operational requirements; manufacturers will regard buildings as essential plant contributing to the production and accounted for in the same way as plant. Buildings will be increasingly designated as either architectural forms where their primary function is esteem or as working infrastructure in which case they will be designed for assembly and disassembly within a sustainable component environment, often with comparatively short lifespans. Facilities management will continue to grow with companies offering hard and soft facilities management.

The public sector will be slow to differentiate between buildings for esteem and buildings for function. However, the latter will grow through Public Private Partnership schemes.

Procurement strategies

Accelerating Change has moved the industry towards approximately 35% of projects by value procured through collaborative procurement methods due to the high cost of implementing such systems. Driven by larger client organisations which own, operate and offer for rental, the construction industry will move away from competitive tendering on full design information to design and construct services utilising structured supply chains with limited use of overt competitive tendering. The assumption is that competition between contractors will be determined by the standard of service, competency, identity of the supply chain and provision of the necessary product to add measurable value to the client.

In the public sector more simplified framework agreements within the Public Private Partnership model will develop with fewer operational buildings retained within public ownership. The specification of such operational buildings will increasingly be performance specification based with high levels of stakeholder input. Across the public sector framework agreements will cover the creation of new physical assets and the maintenance of existing physical assets for periods ranging from five to ten years.

In both the public and private sectors the construction brief will become the tender and audit document most probably prepared by an integrated project team involving the client, contractor and tier 1 members of the supply chain.

The structure of the contracting industry

Shares in major quoted contractors will still be seen in the UK as being high risk and poor investments and this will make them attractive to takeovers and mergers resulting in many companies being owned by non-UK based organisations. The exception to this is private limited companies that will expand. There will be a dramatic slowdown of initiatives and companies will settle to operate efficiently within stable selected markets.

The provision of construction on a fee basis for a guaranteed maximum price against a detailed brief within a partnering arrangement will become the norm. Tier 1 members of the supply chain will be involved as key members of the partnering agreement. A change management structure will be agreed at the outset of projects and the dispute climate will become benign. Facilities management will expand.

Small contractors and sub-contractors will become franchise operations of organisations that will take responsibility for their administration, insurance, training requirements and image. These organisations will grow from the consultancies that currently give tendering, contractual and dispute resolution advice to small contractors and will take the registration of competence responsibility away from government. Standardisation will increase with suppliers offering full service to smaller contractors with security backed by the franchising organisation.

Construction consultancy

Construction consultancy will grow with fewer large firms employing more staff at the expense of small private practices employing fewer than 20 which will decrease rapidly. The architectural and engineering profession will split into small concept-only firms accepting no liability for design and large technical design offices offering solutions backed by project based insurance against any failure.

Future of value management under scenario 2

The future of value management and value engineering under scenario 2 is extensive. Those offering built solutions for rent will use it proactively as a better way of doing business. Those contractors tendering for projects where competitive edge relies on best value service will also use value management within the integrated project team to develop the brief and the project, value, risk and change management strategy.

Forward thinking local and central government agencies will incorporate value management within their treasury departments. Treasury departments will track projects for VM and VE using sophisticated intervention criteria. It will be the treasury departments who procure VM services on behalf of service departments for all projects where they consider the application of VM justified. There will be no accounting of 'savings' as in the USA but the recognition that this is a better way of doing business. Government audit functions will use VM and independent teams on a regular basis to audit for best value either on services or for best value under asset management framework agreements. Large private sector clients may begin to adopt this for value for money audits within their own businesses. Value management practitioners will begin to diverge into those that offer a facilitation service predominately and those that offer a service that is more comprehensive and focused around management consultancy – the domain of the large practices. Turf warfare over who 'owns' value management will continue but the IVM will align itself with well established professional associations in construction, becoming a predominately postgraduate or 'prior chartered entry' institution offering enhanced accreditation to value managers working for large organisations. SAVE International and the IVM jointly recognise each other's qualifications and jointly accredit university courses.

Scenario 3: radical change

The most radical change is the development of scenario 2 but within a compressed timescale, say by 2010. Other radical changes likely to be implemented by 2015 are listed as follows.

Construction client attitude towards property

Excluding housing, the majority of property will be rented or leased. The only public or commercial property in owner occupation by 2015 will be those buildings where esteem is the highest factor or a very high factor in the client's value system.

Property owning companies will raise capital through share issue. Specialist fund managers will establish property investment trusts and unitisation. Changes to capital allowances legislation and accounting rules on the presentation of profits will make property rental extremely attractive. Property based unit trusts will become attractive as pension funds.

Public and private sector property dealings will be only distinguishable by purpose, i.e. to facilitate the making of profits or to facilitate the provision of public service.

Procurement strategies

The objectives of Accelerating Change have been achieved by 2007 and at least 50% of projects by value are procured through collaborative working. Larger client organisations which own, operate and offer for rental will become expert clients employing construction and facilities management expertise in house. These companies will establish supply chain links with those organisations that were previously tier 1 sub-contractors with procurement based on full partnering.

In the public sector more simplified framework agreements within the Public Private Partnership model will develop with fewer operational buildings retained within public ownership. The specification of such operational buildings will increasingly be performance specification based with high levels of stakeholder input.

In both the public and private sectors the construction brief will become the primary organisation document.

The structure of the contracting industry

Major contractors will disappear, with FM firms taking on board capital contracting and design functions.

Supply chain management within a partnering arrangement will become the norm operating within an integrated project, value, risk and change management structure. Small contractors and sub-contractors will become franchise operations of organisations that will take responsibility for their administration, insurance, training requirements and image. Standardisation will increase with suppliers offering full service to smaller contractors.

Construction consultancy

Construction consultancy will grow into property companies or facilities management organisations taking the place previously occupied by the main contractor or become part of an FM company. The architectural and engineering professions will split into small concept-only firms accepting no liability for design and large technical design offices offering solutions backed by project based insurance against any failure.

Future of value management under scenario 3

The future of value management and value engineering under scenario 3 becomes an integrated part of a structured project, value, risk and change management procedure. Identifying, creating, enhancing or advising on best value, team working, the facilitation of value systems and the overt recognition of function becomes the normal way of working.

Forward thinking local and central government agencies will incorporate value management within their treasury departments. It will be the treasury department that procures VM services on behalf of service departments for all projects where they consider the application of VM justified. There will be no accounting of 'savings' as in the USA but the recognition that this is a better way of doing business. Government and large private corporations will use VM and independent teams as an audit function to secure best value on a regular basis.

The VM market will have divided into first, 'cottage industry' practitioners, who are sub-contracted in to co-facilitate the workshop phase for larger VM units or offer services to small client organisations who cannot afford the cost of structured collaborative procurement arrangements but still wish to use VM on their projects; second, VM programme managers who procure VM services from consultants in the marketplace and who will either audit projects and services using independent teams or work with teams of record; third, large consultancy practices with either VM subsidiaries or VM units who offer the service as part of a broader management consultancy service to construction. In the UK there will be a re-alignment of professional associations in construction, with the formal alliance between the IVM, SAVE International and other construction institutes coming into direct competition with other management related institutes with interests in value delivery.

10.5 Achievement of objectives in writing the book

The authors have attempted to bring together and synthesise international developments, benchmarking and an action research programme in VM to provide a comprehensive package of theory and practice. The book has defined the nature of value, transformed it into a set of definitions and also discussed the alignment of VM with TQM. The book has also elucidated the complexity of value systems, and states that they must be addressed in any VM study. The chapters dealing with the project value chain and the value thread have provided the theoretical background to value management. Operationalisation of theory has occurred through the discussions on VM study styles describing inputs, processes and outputs. An enhanced VM process has also been presented together with how that can be adapted to the public sector to deliver best value.

The continued professional development of value management has also been explored through three scenarios indicating that its future may lie with extending the service more into managing value as part of broader management consultancy offered to large corporations and the treasury departments of local and central government agencies or to assisting integrated supply chains. This exploration has suggested that the drivers of corporate patronage will pull VM practice into large consultancy firms serving large organisations on the one hand, and consequently increase the pull towards a 'cottage industry' practice sector on the other where sole traders are sub-contracted in to the larger consultancy firms or act as independent consultants to small clients. The consequence is that VM practice will polarise into those that practice and those that manage VM programmes within public and private sector corporations. The exploration of the scenarios also suggests that the institutional structure for the continued development of VM remains unclear.

The book has also attempted to raise issues for debate and enrich the further development of VM. We hope it has achieved this.

10.6 Reference

1. Strategic Forum for Construction (2002) *Accelerating Change*. Rethinking Construction, London.

Appendix 1
Toolbox

Introduction

The Toolbox describes tools and techniques commonly used in value management workshops. The tools are listed in alphabetical order and are referred to in the chapters to this book. Primary reference should be made to Chapter 5, *Current Study Styles and the Value Process*.

ACID test

Selecting the Team: the ACID test

The facilitator can work with the client in selecting the members of the value management team or can request that the client build the membership of the team based on the inclusion of those stakeholders with an input relevant to the particular stage in the development of the project. The ACID test as shown in Table A1.1 is used to determine who should be a member of the team. Generally

Table A1.1 The ACID test for team membership.

A	*Authorise:*	Include those who have the authority to take decisions appropriate to the stage of the development of the project. Those who have executive authority to take decisions are invaluable members of the value management team through their ability to immediately sanction a particular line of discovery or take a decision during the workshop which resolves an issue or unblocks a particular line of investigation.
C	*Consult:*	Include experts who have to be consulted regarding particular aspects of the project during its evolution at the workshop. If a particular line of investigation is dependent upon consultation with an absent expert workshop progress may be compromised.
I	*Inform:*	Do not include those who have only to be informed of decisions reached during the workshop.
D	*Do:*	Include those who are to carry out major tasks specified at the workshop. In this way those who are, for example, to design or construct, based on decisions taken at the workshop will be fully conversant of that decision.

team membership tends to be greater in number at the strategic stage of projects when a large number of issues are being considered and smaller when the technical details of the project are being investigated.

Factors that the facilitator may wish to take into account in selecting the team are:

(1) Limit multiple representations from one organisation or one department. For example, three members from one organisation where other organisations have single representation will lead to a weight of argument in favour of the multi-represented organisation.
(2) Understand the hierarchical mix (senior, subordinate) within the team.
(3) Understand the relationships between team members. For example, one member may be dependent financially on another member of the team.
(4) Consider the completeness of the team. Discuss with the client any apparently missing members.

Action plan

Actions will always arise out of workshops. The action plan is the summary document that is usually incorporated or appended to the executive summary of the workshop report and describes the action in detail, the members of the team best suited to take the action (whether present at the workshop or not) and the date by which the action is to be completed. Members of the workshop team will carry out all of the items in the action plan if the ACID test was correctly carried out.

The action plan is included in the project execution plan. The team's actions will be reviewed at a future design team meeting.

Audit

An audit is a systematic check or assessment typically carried out by an independent assessor. In the context of a value management workshop the term audit is used to describe the activity of checking stages between workshops. For example, it is entirely practical to undertake the production of a project brief as a performance specification within a value management workshop. However, there comes a point when each member of the design team has to work alone to translate the performance specification into a constructible design. Once this design work is complete an audit is undertaken to determine how effectively the resultant design answers the performance specification. In this context the audit questions are:

(1) Which element of design is being subjected to audit?
(2) What are the strengths of the design in meeting the performance criteria?
(3) What are the weaknesses of the design in meeting the performance criteria?
(4) How might the strengths be improved and/or the weaknesses be addressed?
(5) Do the ideas stimulated by (4) above still meet the original performance specification criteria?
(6) If the check reveals or inspires an excellent design then a decision is taken to move to the next stage.

It is vital that an audit is undertaken at every stage of design development to prevent antipathy towards change which arises from too much work being done and too much time expended before the audit is undertaken. It is vital that the design team accept that this is good practice rather than seeing the audit as a criticism of their work.

Checklist

This technique is adapted from Morris & Hough[1] who undertook a study of major projects internationally. It has been determined that most information relevant to projects can be summarised under the following generic headings. The use of these headings facilitates the discovery of information through either interviewing, interrogation of the team, or issues analysis. These techniques are described elsewhere in the Toolbox.

(1) *Organisation:* The identification of the client's business, the place of the project within the business and the users of the project (who may not necessarily be a part of the client organisation). Under this heading there would be an investigation of the client's hierarchical organisational structure and the client's key activities and processes that would impact the project. Included is information on the departments from which the client representatives will be drawn, the decision making structures of the client and how this will interface with the project design and construction teams, and the communication networks anticipated for controlling the project. These decision making structures become more important in situations where a single project sponsor or project manager represents the client team. The limits to the executive power of the project sponsor or project manager should be clearly defined. Information should be sought about the core and non-core business of the organisation and how they relate to space use if a building project is under study.

It must be recognised that most organisations are dynamic and therefore subject to continual change which should be recognised and recorded.

(2) *Stakeholder analysis:* Following the discussion of the organisational structures it should be possible to identify all those who have a stake in the

project. Stakeholders should be listed and their relative influence assessed.

(3) *Context:* The context of the project should recognise such factors as culture, tradition or social aspects. Cultural aspects may include the relationship of one department with another and the fitting out and general quality of the environment, e.g. a court. Tradition can cover such aspects as corporate identity that may be important in, for example, retailing. Social aspects will generally relate to the provisions made by the client for the workforce: dining, recreation, sports and social club activities, crèche, etc.

(4) *Location:* The location factors will relate to the current site, the proposed sites or the characteristics of a preferred site where the site has not to date been acquired. All projects, whether construction or service, will have a location.

(5) *Community:* It is important to identify the community groups who may require to be consulted with respect to the proposed project. Some market research may need to be undertaken to ascertain local perceptions. The positioning of the project within the local community should also be completely understood.

(6) *Politics:* The political situation in which the project is to be conceived should be fully investigated through the analysis of local government and central government policies and client organisational politics. The latter is often difficult to make overt at a workshop of representatives of different client departments; however, client politics is a key driver behind any project. It is useful to discover the powerful departments, the powerful individuals or groups who are to be represented on the client's project team from which departments they come and how much power they have to make decisions and influence project development. The analysis of political parties in power at national, regional and local levels should include their views on the project, whether there are any changes in political persuasion anticipated over the project life cycle and, if there are, whether these will matter.

(7) *Finance:* The financial structuring of the project should be determined by considering the source of funding, the allocation of funding and the effects of the project cashflow on the cashflow of the client organisation. The latter is particularly important when dealing with public sector organisations working with annual budgets.

(8) *Time:* Under this heading are the general considerations regarding the timing of the project including a chronological list of the procedures which must be observed in order to correctly launch the project. In situations where the project is to be phased, time constraints for each stage of the project should be recorded. This data becomes the basis of the construction of a timeline diagram described later in the Toolbox.

(9) *Legal and contractual issues:* All factors which have a legal bearing on the project are listed under this heading including the extent to which the

client is risk averse and requires cost certainty. Also included here is data relating to the client's partnership agreements with suppliers and contractors.

(10) *Project parameters and constraints:* A primary objective of the value management workshop at the strategic briefing stage is to fix the primary objectives of the project. Therefore it is important that the team understand that the workshop is the end of one stage in the development of the project. Discussions must take place on the evolution of the project to the time of the workshop and to measure the extent to which key stakeholders believe that the project is still evolving. Any constraints surrounding the development of the project should be discussed and recorded.

(11) *Change management:* Value management is a change process which necessarily involves migration from one state to another. The activities involved in change management are evaluating, planning and implementing, usually through education, training, communication, and team and leadership development. As people are at the heart of any change process, communication and involvement are the keys to success. Recognising the change process, the organisation must be able to capture, record, process, structure, store, transform and access information. Change management involves not just managing change within an organisation but also managing risks and anticipating the effects of external factors.

Client value system

For a full discussion of the client value system refer to Chapter 8 *Client Value Systems*.

The client value system diagram (Fig. A1.1) represents the view of the client at the time of the workshop. The stages in the construction of the diagram are:

- Identify the client. The diagram to be constructed represents the views of the client, therefore those constructing the diagram should be members of the client organisation with executive authority and not consultant advisers to the client. The facilitator will ask all members of the team to remain silent while the client completes the diagram.
- Decide the value criteria. The diagram will typically be comprised of up to nine variables. The nine variables described in Chapter 8 and listed in Fig. A1.1 have proved to be the key criteria against which client value relationships can be made explicit. It is likely that additional value criteria will be highly correlated with one of the nine variables so care should be taken in adding further criteria. Ease of maintenance for example will be highly correlated with OPEX and correlated to a lesser degree with environmental impact and comfort. Nevertheless, having explained the nine criteria the

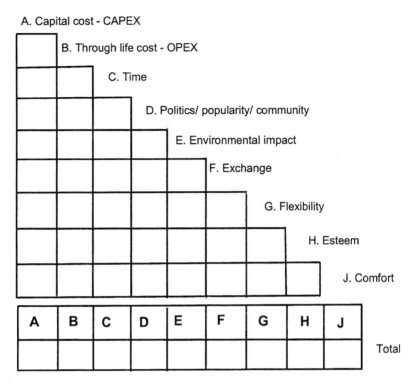

Fig. A1.1 Client value system.

facilitator should explore any other value criteria with the client. Value criteria within the list should be explored for relevance; for example, exchange may not be relevant to a hospital that is to remain in public ownership.

- Organisational or project values? Be very clear as to whether the exercise is relating to the client organisation as a whole or the specific project. In many situations it is necessary to undertake the exercise with the project characteristics set to one side, i.e. to obtain the value criteria for the organisation. Following this the question can be asked, 'How do these vary for the project?' The reason for doing this is that the particular project may have value criteria that would not normally relate to the organisation. For example, a university laboratory may have been destroyed in a fire. The client will be keen to see the laboratory rebuilt as soon as possible. Time is therefore likely to be a key value variable in the context of this project but normally, for the university organisation, time may not be an important value driver. In retailing, however, where the time from obtaining a retail unit to trading is normally kept as short as possible, time might be a key driver for the organisation as a whole.
- Undertake the paired comparison by asking the questions in turn, for

example, 'Which is more important to you, CAPEX or OPEX?' If CAPEX is more important, A is inserted in the appropriate box.
- On completion of the matrix add the total number of As, Bs, etc., and enter the number in the totals box.
- The rank order of the variables will represent the client value criteria; this should be checked back with the client. For example, 'The result based on the first three rankings is that the most important factor to you is esteem, followed by exchange, followed by comfort (attractiveness to occupants).' If the client agrees then the team know that they have to design a landmark building which will attract high value in terms of capital valuation and/or rental levels and will be comfortable and attractive to all those who are to work in the building.
- It is important to note that the numbers in the totals box are ordinal and not interval values. This means that care has to be observed in using the numbers as numbers in later exercises.

Often the client's value system is satisfied by the above exercise and everyone in the team (including the client members) can understand the value importance given to the variables. However, if there appears to be any confusion then each variable can be explored further by examining its continuum. This is done by asking each member of the client organisation who partook in the paired comparison exercise to place a dot on a voting slip which describes each variable. The facilitator will collate the voting slips onto a master diagram (see Fig. A1.2). The clustering of the dots reflects the extent of common understanding between the client representatives present. Note that there are dangers in extending this exercise to include members of the design and construction team; the exercise may just expose the extent to which the design team completely misunderstands

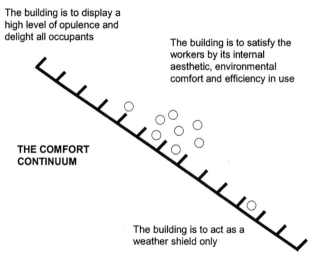

Fig. A1.2 Paired comparison matrix.

the client. It is better that the design and construction team learn from the exercise rather than take part in it.

All the above information is recorded in the workshop report which becomes a key document for audit purposes. An auditor following the decision steps will be clear as to the background of subsequent decisions relying on an understanding of the client's value criteria.

Design to cost – BS EN 12973: 2000 Value Management

Design to cost is a method used from the start of the development programme of a product or system that uses the market price or customer budget as the foundation for production costing. The anticipated production cost is considered as a performance that must be attained together with the technical performance. During development the balance between cost, performance and schedule is continuously assessed. Design to objectives occurs when design to cost expands to incorporate other objectives one of which is cost.

Document analysis

The document analysis serves a similar function to the interviews in informing the facilitator about the background to the project. Whether the document analysis is undertaken first to frame the interview questions or the interviews focus the document search is a decision that can only be taken with reference to the particular project. The document search should initially focus on the client's correspondence file particularly where the project has had a long gestation.

Where a full project execution plan (PEP) is in existence this may negate the need for a document search as all decision taking with regard to the project will be fully recorded. The PEP will include previous VM/VE reports which will be an important source of information.

The facilitator may need to assemble some of the information from the document search for use in the workshop. Certainly, current drawings and contract programme should be available to the workshop team.

Driver analysis

Driver analysis can be undertaken either as a part of the issues analysis or as a self contained exercise. The essential question is 'What/who is driving this project?' It can be utilised in different forms: at project, service or design level. It forces the team to think about what is driving a project or design. At technical project level it can be used to identify what is driving cost, time or quality since these relate to value drivers. At design level is can be used to identify design drivers, for example what is driving:

- Architecture/aesthetics
- M&E
- Structure
- Space

The driver analysis leads towards capturing and encapsulating value drivers, which can subsequently be used to refine function analysis and elements and components function diagramming.

Element function analysis

An element for cost analysis purposes is defined as a component that fulfils a specific function or functions irrespective of its design, specification or construction (BCIS).

An elemental cost plan, a common format in UK, therefore gives the cost breakdown of construction projects according to functions. Theoretically a standard list of functions applies to each element; however, in practice differences occur for each specific building. Table A1.2 (see pages 284–6) is taken from an actual value engineering study and is useful as an aide-memoir. Elemental function analysis is a value engineering technique that seeks to identify innovative technical solutions that retain or add value to the project. Being a value engineering technique the following definitions apply:

Element function analysis seeks to provide the necessary element functions at the required quality and at the lowest cost while identifying and eliminating any unnecessary cost.

The five key questions to ask are:

What is it?	Description of element.
What does it do?	Functional definition of element.
What does it cost?	Exploration of cost to complete value equation.
What else will do it?	Innovative alternatives.
What does that cost?	Comparison of functions given and relative costs.

Facilities walkthrough

This is similar to functional space analysis but relies on following a user's use of the building by following a route on a drawing and asking what function takes place in each space. In situations where the functional use of space is solely for transit a question has to be raised relating to whether the space is unnecessary or whether a user will be wasting time transiting the space for the life of the building.

See also Impact mapping and Functional space analysis.

Failure mode and effects analysis – BS EN 12973: 2000 *Value Management*

Failure mode and effects analysis is a technique used to identify and eliminate possible causes of failure. The technique requires the sequential, disciplined approach to assess systems, products or processes and involves establishing the modes of failure and the effects of failure on a system, product or process. This ensures that all possible failure modes have been fully identified and ranked in order of importance.

FAST diagramming (function analysis system technique)

FAST diagramming is a technique to represent functions diagrammatically and can be used, for example, to illustrate the mission and objectives of a project or the primary function and related functions of a component.

A full description of the technique can be found in Chapter 3.

Functional performance specification – BS EN 12973: 2000 *Value Management*

A functional performance specification is a document by which a customer expresses a requirement in terms of user related functions, constraints and evaluation criteria. The expression of these user needs in the functional performance specification without reference to the technical solution gives freedom to the designer, manufacturer or supplier to select, incorporate or design the most efficient product to satisfy the specification.

Functional space analysis

Functional space analysis relies on a number of sequential techniques using flow diagramming, space descriptions and specifications to accurately describe the functional spaces required of a facility. The functional space analysis forms the basis of the construction of room data sheets and thereby is a foundation component of a construction project brief.

A full description of the technique can be found in Chapter 3.

Table A1.2 Checklist of element functions.

1A Substructure	2A Frame	2B Upper floors
Transmits load Prevents collapse Supports ground floor Minimises movement Retains earth Resists damp Insulates from cold	Transmits load Resists wind load Supports floors Resists excessive deflection Resists fire Resists corrosion Expresses structure Conducts lightning	Transmit load Resist fire Contain space Separate space Receive finish Support load Act as a diaphragm Acoustic barrier Insulate against sound Provide security
2C Roof	2D Stairs	2E External walls
Transmits load Excludes climate Filters climate Filtes sound Resists decay/corrosion Directs rainwater Contributes to aesthetic/built form Creates shading Resists uplift Assures security Attracts lightning	Vertical circulation Connect levels Interrupt floor Transmit load Create means of escape Base for finishes Control safety Aesthetically pleasing	Transfers load Filter climate Resist wind load Resist vandalism Conduct heat Attenuate noise Filter light Prevent spread of fire Support finish Contribute to aesthetic Enclose space Aid security
2F Windows and external doors: section 1 windows	2F Windows and external doors: section 2 doors	2G Internal walls and partitions
Filter light Control ventilation Conduct heat Conduct noise Views (in/out) Create aesthetic Filter climate Ensure security Frequent maintenance	Allow access/egress Aid security Architectural features Conduct heat Control climate Filter light Amplify heat loss Enclose space Control ventilation	Base for finishes Reduce space Divide space Enclose space Attenuate noise Transmit light Support services/fittings Transfer load Maintain security Resist fire Impede ventilation Separate climate Separate function Architectural feature
2H Internal doors		
Enclose space Separate functions Provide access/egress Resist fire Noise reduction Separate environments Support sign Resist vandalism Aid security Aid vision Architectural feature		

3A Wall finishes	3B Floor finishes	3C Ceiling finishes
Modify light Isolate from heat Resist vandalism Spatial awareness Control acoustic Encase services Facilitate upgrading Express function Architectural aesthetic Resist wear Support colour/texture Take up tolerances Allow hygiene Allow security Convey information	Contribute to aesthetic Resist wear Contribute to colour/texture Resist noise transfer Resist slip Enclose services Take up tolerances Respond to function Resist vandalism Respond to life expectation Allow security Control acoustics Spatial awareness Direct traffic	Enclose services Support services Contribute to aesthetic Reduce sound Aid security Create thermal barrier Create plenum Enclose space Aid fire resistance Filter light Reflect light Absorb light Control acoustic Allow flexibility
5 *Services* 5A Sanitary appliances Contribute to aesthetic Soil and waste disposal Ensure hygiene Resist vandalism	5B Services equipment Prepare food Process food Serve food Ensure hygiene Store food Clean equipment	5C Disposal installations Effect disposal Contribute to aesthetic Soil and waste disposal Ensure hygiene Resist vandalism
5D Water installations Distribute hot and cold water Store water Maintain water temperature Contribute to aesthetic Ensure hygiene Resist vandalism	5E Heat source Environmentally friendly Create energy source Heat water Heat air	5F Space heating and air treatment Distribute heat Distribute air Condition air Create noise Minimise heat loss Control internal environment Protect fabric
5G Ventilation Distributes air Removes contamination Creates noise Controls internal environment Conditions air Protects fabric Air for combustion Ensures smoke detection Encourages vermin	5H *Electrical:* Electrical source and mains Distribute power Control consumption Measure consumption Control environment Maintain 24 hr access	5H *Electrical:* Electrical power supplies Drive equipment/plant Contribute to aesthetic

Table A1.2 *continues*

Table A1.2 contd.

5H Electrical: Electrical lighting and fittings	5I Gas installation	5J Lift installation
Light task Emit light Improve security Create noise Maintain fire escape Control energy consumption Control lighting environment Contribute to interior design	Environmentally friendly Creates energy source Distributes fuel Measures consumption	Improves vertical circulation Enables disabled vertical circulation Improves fire fighting Reflects function Improves aesthetics Necessitates regular maintenance Creates vertical structure
5K Protective installation	5L Communications	5M Special: Ramp heating
Assists fire fighting Protects building Protects contents	Distribute information Control information Improve security Protect people Transmit information Aid security Record information Aid disabled people Improve safety	Improves safety Avoids freezing surface Increases drainage
5N Builder's work in connection with services		
Accommodates plant Accomodates distribution systems Penetrates barriers Modifies structural design		
6A Site works	6B Drainage	6C External services
Create access Prevent access Define site Contribute to aesthetic Affect security Modify external spaces Affect maintenance Discourage loitering Resist vandalism	Transmits effluent Controls flow Increases security risk Ensures hygiene Requires access Processes contaminants	Measure water consumption Protect supplies Distribute services Aid security Discourage vandalism

Goal and systems modelling

In its simplest form goal and systems modelling relates the goals of a project with existing systems available to achieve the goal. In its more complex form it draws together the aspirations represented by the function diagram in a quality function definition methodology. A description of this technique is found under QFD below.

An example of a simple system for waste management is shown in Fig. A1.3.

Fig. A1.3 Goal/systems model.

This simple example illustrates the absence of a system for community consultation and minimising environmental impact. The systems that are in place and linked to goals have to be questioned in terms of adequacy.

Idea reduction: judgement

Assume an exercise is being carried out to alleviate station overcrowding. A brainstorming exercise has taken place and the ideas are listed on flip chart paper. Good ideas can be determined in three stages. In the first stage – called 'silence means no' – team members remain silent if they believe that the idea is not worthy of further consideration. The facilitator reads out ideas and if no one has responded within 2 seconds the idea is crossed through. In minutes this can reduce over 100 ideas to a limited few worthy of consideration. The result of the exercise is shown in Fig. A1.4.

In the second stage of the exercise a vote is taken on which ideas are likely to succeed based upon subjective judgement only. An effective method of voting is the sticky dots technique whereby each member of the team is given a number of dots, the number being say 20% of the total number of ideas. The facilitator may ask the team to consider voting only for those ideas which individually they would be willing to champion, whether or not they have the technical expertise to do so. The dots are placed in the column to the right of the descriptions.

In the third stage those ideas which have attracted the most dots are analysed by reference to four criteria:

Idea	Description				
1.	~~Demolish existing station structure~~				
2.	~~Close all retail outlets~~				
3.	Widen platforms for increased capacity				
4.	Lengthen platforms for increased capacity				
5.	~~Reduce number of stopping trains~~				
6.	Increase exits from stations				
7.	~~Make people run~~				
8.	~~Shoot anyone leaving a train~~				
9.	Separate arriving and departing passengers				
10.	Install indicator screens so people wait in the right place				

Fig. A1.4 Initial ideas sort method.

Idea	Description		CA	TF	EV	FS
1.	~~Demolish existing station structure~~					
2.	~~Close all retail outlets~~					
3.	Widen platforms for increased capacity	••••	✓	✗	✓	✓
4.	Lengthen platforms for increased capacity	••				
5.	~~Reduce number of stopping trains~~					
6.	Increase exits from stations	•••••	✓	✓	✓	✓
7.	~~Make people run~~					
8.	~~Shoot anyone leaving a train~~					
9.	Separate arriving and departing passengers	•••	✓	✗	✗	✓
10.	Install indicator screens so people wait in the right place	•••••	✓	✓	✓	✓

Fig. A1.5 Criteria related ideas sort method.

(1) Is the idea acceptable to the client and doers it reflect positively the client's value system? (client acceptable – CA)
(2) Based upon a subjective judgement only, is the idea technically feasible bearing in mind the constraints on the project discovered during the information stage? (technically feasible – TF)
(3) Again based upon subjective judgement only, is the idea economically viable? (economically viable – EV)
(4) Finally, with reference to the function analysis undertaken is the idea functionally suitable? (functionally suitable – FS).

In the example shown in Fig. A1.5 any idea which attracted three or more dots was analysed. In this exercise the two ideas worthy of further development are idea 6 and idea 10.
See also weighting and scoring.

Impact mapping

Impact mapping is a statement of how each individual user has to modify their ideal work method to accommodate the building layout. For example:

- If a user's job requires on average 12 visits per day to a file store that is a 5 minute journey from the user's workstation then that user will spend on average 1 hour per day accessing the file store.
- Sightlines from a nurses' station in a hospital might lead to more or fewer staff being required.

Similar to the facilities walkthrough, a study at design stage of the impact of the building on users may lead to efficiencies in design.

Interviews

Interviews prior to the workshop are undertaken to:

(1) Give the facilitator an overview of the strategic and tactical issues surrounding the project and allow the first identification of mismatches.
(2) Enable the facilitator to begin to understand the issues surrounding the project. In this respect the checklist described in this Toolkit provides a sound basis for the development of the interview questions, addressing for example:
  (a) The place of the project within the client's core and non-core business.

(b) The client's hierarchical organisational structure including a diagram of departmental structure.
(c) The client's key activities and processes that would impact the project.
(d) The decision making structures of the client including timetables of relevant meetings and boards.
(e) The limits to the executive power of the client's project sponsor or project manager.
(f) The identity of all stakeholders.
(g) Tradition and cultural background.
(h) The characteristics of the chosen or yet to be chosen site.
(i) The identity of community groups who may require to be consulted with respect to the proposed project. A general view on whether the project is going to be popular with the project's neighbours.
(j) The impact of a change in the political party in power at local or central government level.
(k) The financial structuring of the project, cash flow issues and annuality.
(l) The timing of the project and the chronological procedures.
(m) Any legal issues and procurement preferences.
(n) Any constraints or boundaries which the client wishes to impose and why.
(3) Sensitise the facilitator to any controversial issues or hidden agenda.
(4) Enable the facilitator to discuss with the client the team membership.
(5) Derive an appropriate agenda and the initial selection by the facilitator of the tools and techniques to be used in the workshop.

Issues analysis

Using this approach the facilitator will ask the team for all factors impacting the project. The issues will tend to be uncovered in a relatively random fashion and therefore the facilitator records the issues on repositional notelets. After, typically, about an hour the team will have exhausted all of the issues impacting the project.

The next stage in the process is for the team to sort the repositional notelets under the 11 generic headings (see Checklist, above) supplemented as necessary. The headings are written on card at the top of blank flip chart paper that lines the wall. Where a particular checklist heading has no repositional notelets attached to it, the facilitator will ask for further issues under that heading.

After all the issues have been sorted each member of the team is given approximately 10 black sticky dots to spend on those issues (on repositional notelets, not on the card headers) that the particular team member believes important.

On completion of this exercise each team member is given approximately five red sticky dots to spend on those previously highlighted issues of importance which the team member believes are so important that the project may be compromised unless that issue is resolved at the workshop.

At this stage all of the issues are under headings and effectively ranked by importance based upon the number of sticky dots attached to the repositional notelets. The facilitator then interrogates the team and records in detail on flip chart paper the information behind those issues that have black or red dots attached to them. It is common for some issues to be ranked highly important by the majority of the team and therefore these are investigated in considerable detail. As in method 1 the issues analysis and the ensuing information on key issues is pinned to the wall and surrounds the team for the remainder of the workshop.

A full description of issues analysis is given in Chapter 3.

Kano

The Kano model[2] developed by the Japanese quality guru Dr Noriaki Kano states that maximum quality is realised when targeted characteristics are achieved and the customer is delighted. There are three variables within the model – basic factors, performance factors and delighters – which have a relationship to the presence of characteristics and customer satisfaction. This is illustrated in diagram A1.6.

In the model a basic characteristic is expected to be present; the customer will be dissatisfied if it is absent and only neutral if the characteristic is completely fulfilled. The performance characteristic relates to the essential function: the

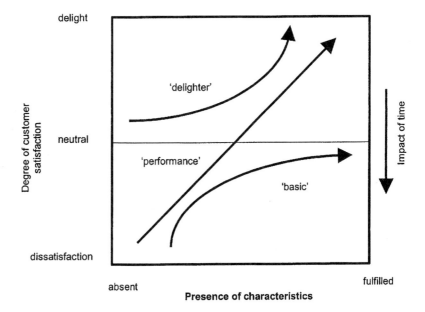

Fig. A1.6 The Kano model.
Source: Adapted from Bicheno (2000) *The Lean Toolbox*[3].

customer will be more satisfied if higher levels of performance are achieved. The delighter is the unexpected extra characteristic. There is, however, a time dimension to the model such that the three variables will tend to sink over time, i.e. what once delighted is now expected and higher levels of performance are always sought. For example, power steering on small cars as a standard feature once delighted customers but now power steering is expected as a basic characteristic and its absence would lead to dissatisfaction.

The Kano model is used in value management to orientate the team following the goal and systems modelling exercise either in its simple form or following a full Quality Function Deployment (QFD) exercise. The construction of the model requires the identification of basic, performance and delighter characteristics. In the waste management example used in the description of the Goal and Systems model above, kerbside waste collection from a single bin is a basic function. Performance characteristics would relate to the number of collections and the build up of waste at customers' premises. Delighters generally refer to innovations. For example, it may be possible to incorporate a deodorising spray to the emptying mechanism on the collection vehicle so that each bin has a pleasant smell at the time of first use. Analysis of basic factors, performance factors and delighters enables the team to be focused in terms of actions that need to be taken forward. A failure of a basic function will lead to dissatisfaction irrespective of the number of delighters.

Lever of value

The lever of value (Fig. A1.7) is an orientation tool. It describes the characteristic stages in the development of a project and illustrates the value potential.

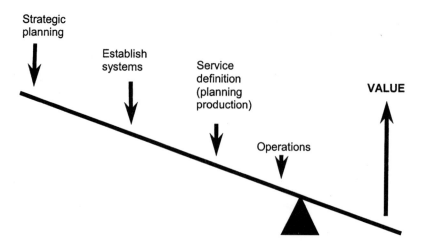

Fig. A1.7 Lever of value.

Life Cycle Costing

See Whole life costing.

Post occupancy evaluation

A fully structured post occupancy evaluation of the client's last completed building can highlight efficiency gains that might be possible in the current project. A note of caution, however, is not to collect information for the sake of collecting information. All information sought should be sought for a reason and be capable of being presented to the individual members of the workshop team in a form conducive to being understood in the shortest possible time.

Steps in a post occupancy evaluation include:

(1) A clear decision on the sort of information to be collected and key questions to be asked of the information. The questions may have arisen from a document search of the proposed project or from interviews.
(2) Examination of all possible documented information including pre and post contract correspondence files, the project brief, drawings and specification, the tender documentation and preferably an elemental cost analysis.
(3) Examination of any previous customer care surveys.
(4) A schedule of areas of functional space.
(5) A walkthrough of the existing building addressing the questions highlighted in (1) above, taking photographs as necessary.
(6) Interviews with users of the existing facility.

Questions that might be posed prior to commencing a post occupancy evaluation might include:

- In comparing or benchmarking costs, how did the last completed building of the client compare with the cost of other similar buildings from the same industry sector?
- Are there any obvious elements of redundancy? For example possibly the structural design incorporates a frame. Is the building sufficiently and permanently compartmentalised such that the use of a frame might be questioned as redundant?
- A common feature of post occupancy evaluations is the fact that considerable complaint attaches itself to items which were promised but which do not appear in the final outcome. Is there evidence of this here?
- Is there evidence that the designers have allowed a particular briefing statement to drive the design. For example, the brief might state 'the design

to provide as much natural light as possible ...'. This could easily drive the whole design approach if not queried.
- What is the percentage of circulation space?
- Are there any specific safety or security issues?
- Is there evidence that a standard specification has been slavishly adhered to? For example, it may be that the hardwood joinery and suspended ceilings are used in all areas including those where they are inappropriate.
- Has sufficient thought been given to cleaning and maintenance?
- Are the heating, lighting and ventilation sufficient and comfortable?
- Are the occupants satisfied? Do they feel that the building works for them or are they constantly adapting their own systems to match the configuration of the building?

Presentation

A presentation is an opening technique when given at the commencement of the workshop to provide key stakeholders with an opportunity to present their viewpoint before the main workshop process commences. Presentations are also useful as a closing down technique when given at the completion of a workshop, both for working groups to present their ideas and solutions for interrogation and validation by other working groups and for decision takers invited to the final stages of the workshop for that purpose.

Presentations can also be used as both an opening and a closing technique as part of the workshop process during plenary sessions. As an opening technique the presentation is being used to inform other groups of ideas and current thinking for further discussion. As a closing technique at a final plenary session ideas can be presented back for agreement.

Process flowcharting

See Functional space analysis.

Project drivers

Projects have a wide variety of drivers – that is, those factors that are giving the project momentum. As a part of the issues analysis it is often beneficial to attempt to identify these drivers to determine whether they are strategic, legislative, organisational or planning drivers, or technical drivers. Technical drivers may be a result of the one or more of the other drivers but unless this fact is realised they are able to assume a life of their own.

Project execution plan (PEP)

The project execution plan is a dynamic document that commences at the inception of the project and includes, where appropriate, the options appraisal report. It is a dynamic management document used by all members of the team that records the project strategy, organisation, control procedures and responsibilities. It contains a formal statement of:

- The user needs (the strategic brief).
- A performance statement of all aspects of the project (the project brief).
- The strategy agreed for their attainment (the project execution strategy).

It is a live, active management document, updated regularly during the project's life cycle and used by all parties both as a means of communication and as a control and performance measurement tool. It begins life as an empty file with dividers indicating the documents to be included.

Examples of the items that a PEP should contain are:

- The project mission.
- The aims and objectives of the PEP.
- The procedures for updating the PEP.
- The project organisation structure of the client.
- A list of consultants and a description of the consultants' responsibilities.
- The contractor's, management contractor's and construction manager's organisation.
- The form of contract, partnering agreement, etc.
- Project reporting procedures and particularly the procedures for information distribution and communication between the client's project team, consultants and contractor.
- A full project brief incorporating the client's value system.
- The health and safety plan.
- The quality plan.
- The latest cost report and predicted cost and cash flow. This will be updated at regular intervals.
- The executive summaries and action plan from value management workshop reports.
- The risk management strategy and the latest risk analysis.
- Copies of key permission documents such as the planning permission, the building warrant, listed building consent, tree preservation orders, etc.
- The latest contract programme with milestone activities.
- Change management procedures and design freeze dates related to milestone activities.
- A schedule of key meetings and workshops (including value and risk workshops).
- Procedures for PR and dealing with community and media enquires.

296 Appendix 1

Quality function deployment

Quality function deployment (QFD) as a technique originated in 1972 at Mitsubishi's Kobe shipyard. The technique aims to represent on a single diagram (the House of Quality) the relationship between functional requirements and technical solutions. In addition to its traditional application in manufacturing the technique can be adapted to offer valuable insights into a wide variety of construction management and service applications. In a value management context the use of the House of Quality is preceded by function analysis and goal and systems modelling.

In the context of value management QFD is a team orientated tool that aims to promote the exploration of innovative ways of meeting the necessary functions represented on the function diagram. It does this by translating functions into measurable goals, thereby promoting the identification of the optimal service or product solution. The measurable goals are themselves attributed to stages of the lever of value, namely strategic, systems, service definition (production planning) and operations. The aim of QFD is to represent all attributes on a single diagram and thereby ensure a complete understanding. In order to construct the QFD diagram it is necessary to complete the function diagram and the goal and systems model above. An innovation session precedes the discovery of missing systems or attributes. In Fig. A1.8 two level 2 functions are required but have no systems in place to satisfy them; one level 1 function is only partially satisfied by system 3.

Once assured that systems are in place to satisfy all of the required goals the house of quality diagram is constructed as shown in Fig. A1.9. The phases of construction are:

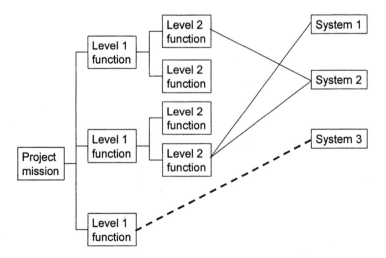

Fig. A1.8 Function diagram.

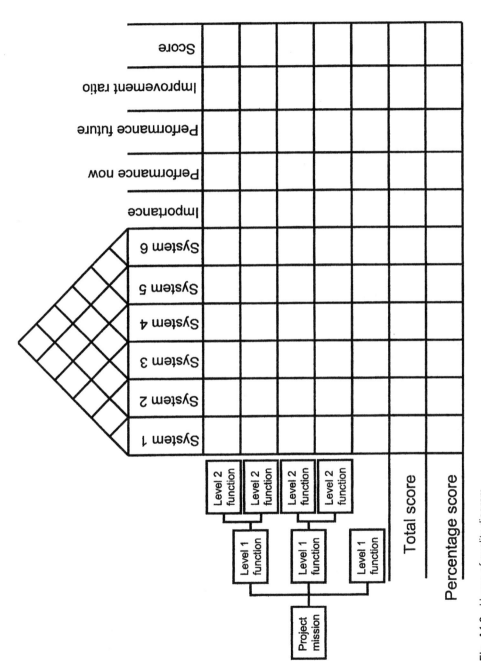

Fig. A1.9 House of quality diagram.

(1) The function diagram on the left of the diagram is constructed or alternatively the functions from the extreme right of the function diagram are listed. These represent the goals, often referred to as 'the voice of the customer'.
(2) Across the top are set out the systems which are either in place or have been derived as a part of the innovation exercise.
(3) The roof of the house accommodates the relationships between systems.
(4) The relationship between the goals and the system is assessed to complete the matrix, the wall of the house. Traditionally a double circle is entered in the cell denoting a strong relationship and is usually given a score of 9, a single circle indicates a moderate relationship and is given a score of 3 and a triangle indicates a possible or low relationship and given a score of 1. A blank cell indicates no relationship as has the value 0.
(5) The roof of the house is also completed using the same three symbols. This denotes the relationship between systems.
(6) The right-hand side of the diagram is a facility to undertake weighting and scoring exercises as follows:
   (a) Column 1 defines the importance of the identified goals using a rating scale of 1–5 where 5 represents very important and 1 signifies low importance.
   (b) Column 2 rates how well the client's service satisfies the identified goals based upon the judgement of the team . The rating scale is again 1–5 where 5 represents excellent service, 2 the lowest level of service to perform the basic function and 1 where the function is not provided.
   (c) Column 3 is the team's opinion of the possible future provision. This may be based on the team's estimation with or without formal benchmarking with other providers of similar services. The scoring system is as column 2.
   (d) Column 4 is the improvement ratio computed by dividing the future rating by the current rating.
   (e) Column 5 is the score for each function determined by multiplying the goal importance by the improvement potential and representing this as a percentage such that the summation of all of the scores equals 100.
(7) The total score for each system is calculated by first multiplying the values for each goal (9, 3, 1 or 0) by the percentage score for that goal and then summing through all the values for that system. The total score for the system can be converted to a percentage such that the addition of the scores for all systems equals 100.
(8) The system on which the client should commit the majority of resources is indicated by the highest system score.

REDReSS

Redress is an acronym for the final stage of the information validation exercise. Key prompts allow a final analysis of the information to ensure that it precisely represents the project. The prompts are:

Re-organisation
Expansion
Demolition
Refurbishment and Maintenance
Safety
Security

A full discussion of this technique may be found in Chapter 3.

Risk analysis and management

Risk and value analysis/management are often linked at the service level. Risk is commonly defined as being a hazard, the chance of a bad consequence or loss, or the exposure to mischance. However it is defined it is normally considered to be those issues that prejudice the outcome of an event. Risk management is a planned and systematic process of identifying, analysing and controlling the outcome of a particular event to achieve the planned objective and thereby maximise value in the project proess. It is at this level that value and risk intertwine using a common team, workshop structures and techniques. It is recommended, however, that value and risk are not so intertwined that the team is constantly flipping from value to risk to value but rather that a value workshop is completed to the evaluation stage before addressing the risks associated with each of the evaluated ideas. The designing out or minimisation of risk will take contingency from the cost and/or time side of the time, cost and quality equation thereby providing, at the very least, identical quality for less money and/or time.

Risk management incorporates three distinct stages:

- Risk identification.
- Risk analysis.
- Risk response.

Risk identification

Risk identification can be undertaken by a facilitated team brainstorming risks associated with a particular solution evolved during the evaluation stage of a value management workshop. Generally these risks will fall under four headings:

- *Changes in project focus:* This will result in a change to the project mission and could result from, for example, a reorientation of the client's core business.
- *Client changes:* These are brought about by unforeseen changes in the client organisation. However, client changes commonly result from poor communication structures leading to incorrect briefing of the project in the first place. The client change is not so much a change as a correction to the project's course.
- *Design changes:* These result from an incorrect analysis of data or the exposure of some unforeseen circumstance, for example forming a doorway in a wall which was assumed to be plasterboard on a timber stud frame but which when the plasterboard is removed turns out to be plasterboard dry lining to a reinforced concrete structural wall.
- *Changes in the project environment:* These changes are brought about by bad weather, non-delivery of materials, unavailability of labour, new legislation, planning restrictions, etc.

Upon completion of risk brainstorming some form of ranking exercise is undertaken to highlight, based upon the opinion of the team, those risks which have a high probability and a serious consequence. The consequence can be determined at three levels; irritating background noise, turbulence in the project's progress (the project can continue but is severely disrupted) or a blocking force which is capable of halting the project until contained. A suitable method is to rank each risk on a scale of A to F as illustrated in Table A1.3.

Table A1.3 Risk ranking.

Rank	Probability	Consequence
A	High	blocking force
B	Low	blocking force
C	High	turbulence
D	Low	turbulence
E	High	noise
F	Low	noise

Those risks ranked D, E and F are examined to make an immediate assessment of whether further action is necessary, in which case they proceed to the next stage, or whether the risks can be taken on board during the development of the project. Risks ranked A, B and C are automatically taken forward to the next stage, risk analysis.

Risk analysis

The first stage in a risk analysis is normally qualitative. The following are analysed:

- A brief description of the risk, the stage of the project when it could occur and the ownership of the risk.
- The factors that could cause it and the likelihood of those occurring.
- The extent to which the project will be affected.

Even where the qualitative risk analysis is considered sufficient the action of undertaking it will sensitise the team towards the recognition of the risk and prompt an appropriate risk response in the event that it occurs. The team, however, may decide that a qualitative risk assessment is insufficient and require a quantitative risk assessment. This is an activity normally not carried out within the workshop and it may therefore be necessary to adjourn the workshop at this point.

Quantitative risk analysis seeks to mathematically model the probability of the risk occurring in two ways: objective risk analysis and subjective risk analysis. An objective risk is when the probability is known exactly, for example the loss of £10 relies on the flip of a coin landing tails up. The probability of this is 50%. A subjective risk is when the probability is not known exactly but can be estimated, for example a loss of £10 relies on more than one hour of continuous rain next Thursday. While reference to weather data records will allow an assessment of the probability of continuous rain next Thursday this cannot be relied on exactly. Quantitative risk analysis becomes mathematically complex when a number of risks are combined. Computer software is available to calculate probability curves for this situation, usually based on a simulation. It should be emphasised that that the results presented by the computer software are merely an aid to decision taking.

Risk response

At the end of the risk analysis exercise the team will undertake a risk response by reducing the cause of risk in one of four ways:

(1) To avoid the risk by undertaking that part of the project in a different manner.
(2) To reduce the risk by taking action to lower the probability of the risk occurring.
(3) To transfer the risk to a third party, commonly an insurance company.
(4) To accept the risk and manage its consequences. This is a valid course of action. If the risk event were to arise the team is sensitised to its recognition and mentally prepared for some form of action.

This stage is characterised by continual reference back to the function analysis and particularly the client's value system. All decisions must accord with the functional requirements and fulfil the requirements of the client. This is a vital part of workshop recap and audit.

Risk register and action planning

The output of a value and risk management workshop is a risk register that summarises the deliberations of the team and records:

- A description of the risk.
- The impact of the risk and the probability of its occurrence.
- The nature of the solution agreed by the team.
- The person responsible for action to the next stage.
- The time or cost contingency which is to be built into the project at this stage.

Action planning requires the team to select the best value for money solution from the evaluation stage and decide:

- Who is responsible for taking action.
- By when is the action to be taken.

Site tour

For the facilitator and the major participants to visit the site is a major advantage in placing the project within its physical confines. Where a site tour is not possible or where the site has only been visited by a number of workshop participants, photographs may be sufficient to answer any queries.

SMART methodology

SMART is an acronym, in a value management context, for simple multi-attribute rating technique. Green[4] advocates the SMART methodology for any design process that has multiple objectives, and most do. The methodology relies upon the construction of a tree diagram, similar to FAST, which represents a hierarchy of design objectives. The highest order objective at the left-hand side describes the resultant and the branches to the left the means to achieving the resultant. The primary difference between a FAST diagram and a SMART diagram is that a FAST diagram is project function orientated with the mission of the project contained in a statement on the left of the diagram whereas SMART is object orientated with the resultant describing the object of the design.

SMART also has a focus on decision support and uses weighting and scoring systems to assist teams in reaching an appropriate decision.

For example, taking the subject of the case study in Appendix 2 – an area community office for a local authority – a SMART diagram produced in an iterative and discursive process is given in Fig. A1.10. The diagram represents the design objectives with the community office being the resultant descriptor. In contrast, a FAST diagram seeks the mission of the project in functional terms with the branches of the diagram representing the strategic functional needs and wants without referring to the design or indeed any technical provision.

The power of the SMART technique comes from the stage following the completion of the diagram where a numerical value is attributed to each level of the diagram. At level 1, the branches of the diagram, there are four objectives. The team is asked to decide on an importance weighting for each objective such that the summation of the values equals 1. The values are shown in Fig. A1.11. In the example the team have decided that 'flexible space for multi use' is of primary importance and have weighted this at 0.40.

The next stage is to take each of the level 2 clusters in turn and carry out a similar weighting exercise. A level 2 cluster is the twigs on each branch, for example, 'flexible space for multi use' has three twigs. In the weighting exercise each cluster must sum to 1.

Multiplying level 1 by level 2 gives a product value which summed through the whole diagram will equal 1. For example, the product of 'flexible space for multi use' (value 0.40) and 'maximum community use' (value 0.50) gives a product value for 'maximum community use' in the context of the area community office of 0.20.

SMART therefore demonstrates to the team the relative emphasis to be given to the various design objectives. It may have come as a surprise that the stakeholders to this project have rated 'maximum community use' and 'day nursery facilities' as the two highest objectives pushing 'improving local authority response time' – previously thought to be the prime objective – into third place. The diagram becomes a catalyst for any ensuing discussion.

It might be argued that the weighting and scoring exercise is highly subjective and that the multiplication process may lead to distortion. This argument can be countered by undertaking sensitivity studies, in this case changing the value of the weighting given. For example, if 'flexible space for multi use' (value 0.40) were to be reduced to 0.35 and 'local presence of local authority departments' were to be increased to 0.30 then what would be the impact on the relative positions of the level 2 objectives? In this case 'maximum community use' (new value 0.175) and 'day nursery facilities' (new value 0.14) retain their position as the two highest objectives with 'improving local authority response time' (new value 0.135) still in third position albeit with a decreased margin. This may assist the decision process.

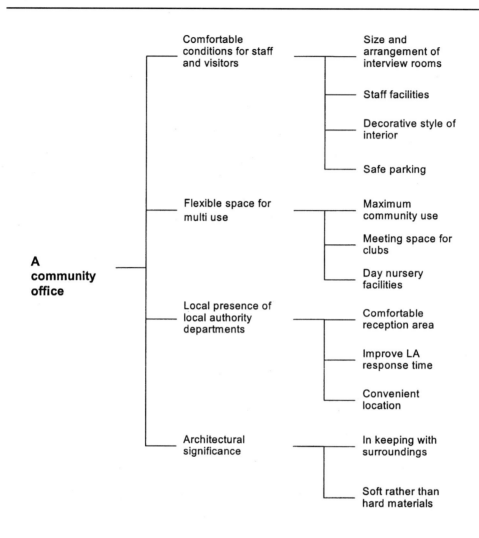

Fig. A1.10 A SMART diagram of an area community office.

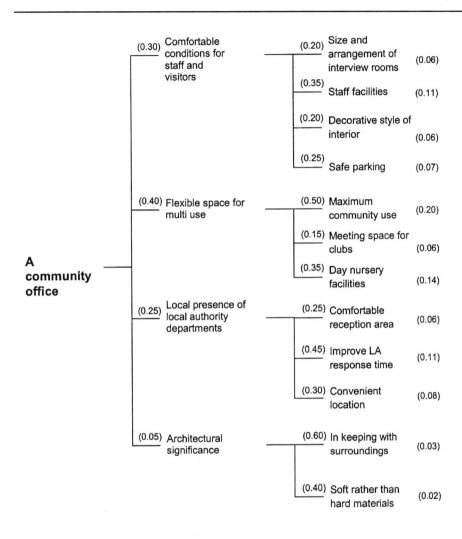

Level 0 **Level 1** **Level 2**

Fig. A1.11 A SMART diagram with decision weights applied.

Spatial adjacency diagramming

Spatial adjacency diagramming can be used as a briefing tool or as an audit tool to analyse a drawn layout. The procedure, as described in Chapter 3, involves identifying each space with a distinct name. These names are transferred to the adjacency matrix diagram (Fig. A1.12). The adjacency requirement between

Courtroom

5	Judge's suite				
2	1	Solicitors' library & robing room			
5	-1	2	Police office & cells		
1	-1	4	5	Prisoner interview room	
5	0	-1	-5	-5	Jury room

Fig. A1.12 Adjacency matrix.

each space is determined on an index scale of +5 to −5. In this context +5 adjacency means that there is a physical link between one space and another whereas −5 adjacency means that spaces are completely separate from one another in terms of environment, sound and physical linkage.

Strategies, programmes and projects

On completion of the information stage it should be clear whether the exercise on which the team is engaged is a strategic development required by the client, a series of programmes to meet the strategy, a series of projects to meet the programme or a combination of all three. It is not unusual to have commenced a workshop thinking that the team was working on a project when it was in effect a programme of linked projects. Figure A1.13 gives focus to the team's thinking.

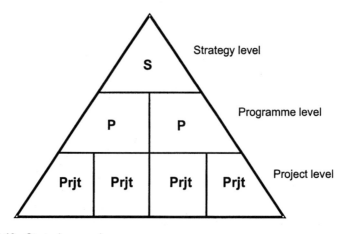

Fig. A1.13 Strategic cascade.

Strengths, weaknesses, opportunities and threats (SWOT)

This is a general management technique but one that can be used at the information stage to scrutinise a service or a design by analysing its strengths and weaknesses, the opportunities for improvement and the threats imposed by adopting a particular approach.

Time, cost and quality

A triangle is drawn on a flip chart in front of the team (Fig. A1.14). The team is invited to agree on the position of a dot within the triangle that describes the relative importance of the parameters of time, cost and quality in relation to the project. A dot hard against the time corner would indicate that time was all-important to the extent that the client would accept an increase in cost and the lowering of quality. A dot hard against the cost corner would indicate that the project had to come in on budget even if time was exceeded and quality was lowered. Finally a dot in the quality corner would indicate that a stated level of quality has to be achieved even if cost and time are exceeded.

This simple technique can easily lead to considerable debate. Once consensus is reached the diagram is again pinned to the wall in full view of the team for the remainder of the workshop.

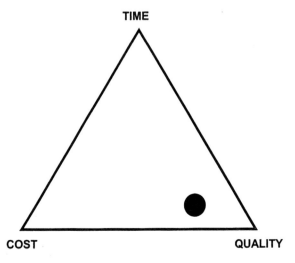

Fig. A1.14 Example time, cost, quality triangle.

Timeline

On flip chart paper in front of the team horizontal lines are drawn to represent the number of years over which the project may extend. The date of the workshop is indicated. All of the preparatory stages of the project up to the decision to proceed are included on the timeline as are key events during the progress of the project prior to its adoption by the client's core business team. This is a useful technique that focuses the team on the permissions and procedures that precede the decision to proceed with the project and the major events during the project. The timeline should not be confused with a highly developed computer generated programme. The timeline is purely a focusing technique used during the information stage.

User flow diagramming

See Function space diagramming in Chapter 3.

Value analysis – BS EN 12973: 2000 *Value Management*

Value analysis is defined as an organised and creative approach using a functional and economic design process that aims at increasing the value of a product subject. Function analysis involves identifying functions and validating and characterising them with the help of clear logical statements, an approach that enhances the communication of a common understanding between team members as to the project's fundamentals. Function cost is the total of the expenditure (use of resources) predicted or incurred in providing a function in a product. The sum of the costs of providing all the required functions should equal the total cost of the product.

Weighting and scoring

Weighting and scoring techniques are relevant in value management and particularly value engineering exercises where a decision needs to be made in selecting an option from a number of competing options and the best option is not immediately identifiable.

Weighting and scoring lies at the heart of many decision support systems including the technique of multi-attribute value theory which itself is related to value trees and SMART (see above).

The first stage in the weighting and scoring methodology is to determine the criteria by which the options are to be judged. In selecting criteria it is important not to select criteria which are highly correlated, for example in judging between floor finishes it would be a mistake to include criteria such as ease of cleaning as well as cost of cleaning since the two are highly correlated.

Consider the following example. A large research consultancy organisation undertakes research, development and training for a wide variety of public and private sector organisations. It is currently commissioning a 12-storey 8000 square metre building on a city centre site. Its work is organised around projects for specific clients that tend to last for between two and six years. Teams for each project will have a dedicated space for the period of the project. Residual space is rented on short leases at attractive rates. The building interior has to be flexible to cater for reorganisation on a two to six year basis. In considering internal partitions a number of options have been suggested. In determining the criteria for judging the options the following have been agreed:

- The ability to be demounted easily with minimum disruption to services, structure and finishes.
- Good noise attenuation.
- Attractive finish.
- Ability to conceal services.
- Ability to support fittings and fixtures.
- Cost.
- Reliability of supply over a period of years.

A paired comparison exercise is held to determine the weighting to be given to each attribute as shown in Fig. A1.15. The weights are carried forward to the scoring matrix (Fig. A1.16) and entered under their respective attributes. The scoring exercise then determines how well each option meets the attributes based on a scale of 1 to 5. These scores are entered in the top left triangle in each cell of the matrix. The score is multiplied by the weight in each cell and the amount entered in the bottom right triangle. All amounts are summed for each option and the total entered. Based upon the decisions taken by the team, traditional stud and plasterboard partition is the best option with traditional plastered blockwork a close second. The propriety partitions did not score well in the exercise and this may require a second look. Indeed a sensitivity analysis should be carried out by changing some of the weights and some of the scores to see the impact. In the exercise shown in Figs A1.15 and A1.16 the team gave a high weight to sound attenuation against which some of the proprietary partitions scored poorly.

See also Idea reduction: judgement.

Appendix 1

```
          A. Demountable
    ┌───┐
    │ B │ B. Noise attenuation
    ├───┼───┐
    │ C │ B │ C. Attractive finish
    ├───┼───┼───┐
    │ A │ B │ C │ D. Support fittings
    ├───┼───┼───┼───┐
    │ A │ B │ E │ E │ E. Conceal services
    ├───┼───┼───┼───┼───┐
    │ A │ B │ F │ F │ E │ F. Capital cost
    ├───┼───┼───┼───┼───┼───┐
    │ A │ B │ G │ G │ E │ G │ G. Maintenance cost
    ├───┼───┼───┼───┼───┼───┼───┐
    │ H │ H │ C │ H │ E │ F │ G │ H. Reliability of supply
    └───┴───┴───┴───┴───┴───┴───┘
```

A	B	C	D	E	F	G	H
4	6	3	0	5	3	4	3

Fig. A1.15 Paired comparison.

Proprietary:

	Demountable A (4)	Noise attenuation B (6)	Attractive finish C (3)	Support fittings D (0)	Conceal services E (5)	Capital cost F (3)	Maintenance cost G (4)	Reliability of supply H (3)	Total
Blockwork	1 / 4	5 / 30	4 / 12	4 / 0	3 / 15	4 / 12	4 / 16	5 / 15	104
Stud	3 / 12	3 / 18	4 / 12	3 / 0	5 / 25	5 / 15	3 / 12	5 / 15	109
Metal	5 / 20	3 / 18	3 / 9	1 / 0	2 / 10	2 / 6	4 / 16	2 / 6	85
Timber	5 / 20	2 / 12	4 / 12	1 / 0	2 / 10	3 / 9	4 / 16	2 / 6	85
Polystyrene	5 / 20	1 / 6	2 / 6	1 / 0	2 / 10	4 / 12	4 / 16	2 / 6	76
Plasterboard	5 / 20	1 / 6	4 / 12	1 / 0	2 / 10	5 / 15	4 / 16	4 / 12	91

Fig. A1.16 Weighted scoring matrix.

Whole life costing

Whole life costing is a technique for economic evaluation which accounts for all relevant costs during the investor's time horizon, adjusting for the time value of money. It is a methodology for predicting present and future costs for the purposes of comparing options and/or determining the most probable future facilities management cost of a facility. A fundamental principle of value management is that options generated through innovation are evaluated using whole life cost criteria. It is also a principle of value management that the cost effectiveness equation takes account of the whole value of the project to the client. For example, the effectiveness of a supermarket may be judged in terms of the number of customers who can park their cars, move efficiently through the shop, pass through the checkout without waiting, return to their car and remove it to allow space for the next customer. This equation is more complex than just providing the lowest cost building.

Relevant costs

A whole life cost calculation assumes an estimation of all relevant costs. These are:

- The investment costs including site cost, design team fees, bidding costs, legal fees, construction cost, tax charges and allowances (capital equipment allowance, capital gains, corporation tax, etc.) and development grants.
- The facilities management costs including:
    - Energy costs (heating, lighting, air conditioning, lifts, etc.).
    - Non-energy operation and maintenance costs, for example letting fees, maintenance (cleaning and servicing), repair (unplanned replacement of components), caretaker, security and doormen, insurances and rates.
    - Replacement of components (planned replacement at end of useful life).
    - Residual credits normally described as those amounts remaining at the end of the investor's time horizon and terminal credits being those values remaining at the end of the useful life of the component or facility. Terminal credits generally refer to the value of a failed component when offered in part exchange or to the scrap value of the component. In determining these credits in the context of a whole building it is necessary to separate the value of the building from the value of the land. Generally the development potential of the land appreciates in value but buildings depreciate until they become either economically or structurally redundant.

There are also costs that are not relevant and are not accounted for in the calculation. These costs are:

- Trivial in amount.
- 'Sunk costs', i.e. those costs relating to the project that the client has already incurred and expended.
- Costs that remain unchanged in comparing options. For example, when comparing double glazing with single glazing there is no need to take account of the window cleaning costs. Similarly, if property tax or insurance were assessed on an area basis they would be excluded from a comparative cost calculation unless comparing solutions of differing areas.

The investor's time horizon

Simplistically, an investor in this context can be regarded as the person or organisation commissioning the whole life cost study. The investor's time horizon is therefore the period of time over which the investor has an interest in the project or the project's sub-systems. The investor may require a prediction of the whole life cost of a complete public private partnership project or may require a comparative calculation on three insulation options. The investor's time horizon therefore the may be judged in terms of the time from the present until the predicted time of sale, redevelopment, demolition or disposal. To accurately determine the time element of the whole life cost calculation is important but is rarely difficult.

Adjusting for the time value of money: present value

Present value may usefully be considered to be the amount of money invested in the bank today to pay for all relevant costs at a given interest rate over a given period of time. The sum to be invested will be less than the sum of all of the costs because a proportion of the whole life costs will occur in the future and therefore the sum invested today will attract interest until the time when it is spent. For example, if a bank was offering 4% interest then how much would need to be invested in the bank today to replace a component currently costing £6000 and estimated to fail in 10 years' time? For the purpose of this example it is assumed that there is no inflation and therefore the component will be the same price in 10 years' time as it is currently. The sum to be invested can be determined by reference to the present value of £1 table in any book of valuation and conversion tables.

Rate per cent = 4   Years = 10   Factor = 0.6755642

Present value: £6000 × 0.6755642 = £4053.39

Therefore £4053.39 is to be invested today to pay for a £6000 component in 10 years' time.

At the time that valuation and conversion tables were first produced hand-held calculators had not been invented. Today any whole life cost calculation can be undertaken with an inexpensive hand-held calculator; further, it is likely that the calculator will cost less than the book of tables.

To calculate a present value use the equation:

$$P = \frac{A}{(1+i)^n}$$

Where:

P = present value
i = interest expressed as a decimal rate per time period – usually years
n = number of time periods – usually years
A = the future amount

In the example above

$$P = \frac{6000}{(1+0.04)^{10}}$$

$$P = 4053.39$$

This calculation ignores inflation which in the real world cannot be ignored. Assuming that the bank base rate is 4% and that inflation is 2% then the accurate discount rate that takes account of the effect of inflation is given by the formula:

$$i = \frac{(1 + bankbaserate)}{(1 + inflationrate)} - 1$$

$$i = \frac{(1.04)}{(1.02)} - 1$$

$$i = 1.96\%$$

Example: If the bank base rate is 4% and the inflation rate is 2% then what is the present value of a £6000 component in 10 years time?

$$P = \frac{A}{(1+i)^n}$$

$$P = \frac{6000}{1.0196^{10}}$$

$$P = 4941.43$$

Simplistically, £4941.43 should be invested in the bank today at 4% interest to purchase a component which currently costs £6000, but in 10 years' time will cost £7314 if inflation is at 2%.

Appendix 1 references

1. Morris, P. W. G. & Hough, G. H. (1987) *The Anatomy of Major Projects: A Study of the Reality of Project Management.* Wiley, Chichester.
2, 3. Bicheno, J. (2000) *The Lean Toolbox*, 2nd edn. PICSIE Books, Buckingham.
4. Green, S. D. (1992) *A SMART Methodology for Value Management.* Occasional Paper No. 53. Chartered Institute of Building, Ascot, Berkshire.

Appendix 2
Case Study

Introduction

The primary purpose of the case study presented here is to demonstrate the tools and techniques of value management. The case study has been significantly changed from the real project that spawned it. The modifications have been designed specifically to highlight factors that occur in real situations. The case study has been used as a vehicle for training and the illustrations given below are from teams involved in a training exercise.

The context of the case study is well described in the client's brief and the architect's proposal. The background to the value management exercise is that following the cost estimates prepared by the quantity surveyor a considerable overspend is anticipated if the project proceeds as planned. The value management exercise has been requested with the primary aim of reducing the overall cost without impacting the quality aspects of the client's requirements.

Earlier text has indicated that this style of value management exercise does not make best use of a process that is better used proactively than, as in this case, reactively. Unfortunately this style of value management exercise is all too often undertaken within the context of exactly the situation described above.

It is acknowledged that some aspects of the architectural and structural engineering design have been deliberately overstated to illustrate particular value management opportunities and the cost plan is intended for illustrative purposes only.

The case study

The case study presented here has been considered in three different ways. First, since the sketch design has been prepared a value engineering approach will be taken. This approach preserves the design as presented and looks to element functions to drive the search for innovative solutions that will provide the function at the required quality and the lowest cost.

Immediately following this elemental approach is an illustration of a study to discover the primary reason for the element being configured as drawn and therefore its primary functions. It should be recognised that backward seeking in this manner may result in radical changes to the design.

The final illustration is of a study that follows the logic of the project from pre-brief through briefing to sketch design. The study demonstrates a number of the tools and techniques described in the text. It is recognised that to undertake a value management study at the completion of a sketch design will result in abortive design work. This may still, however, offer the client the best value for money solution.

Area community office project: brief

Introduction

This project is to provide an area community office for the Council within the Muir housing estate.

The Muir housing estate was built between 1965 and 1971 and is situated on the northern outskirts of the district. The estate as built was a little less than 1 square mile in area and housed 50 000 people in a mixture of high-rise and four-storey maisonettes with several clusters of old people's bungalows. Many of those who moved into the estate were rehoused from slum clearance programmes undertaken in or near to the town centre.

The Muir housing estate is an area of high unemployment with many examples of social deprivation. There have been a number of initiatives by the Council to alleviate the problems of the estate including an innovative storey reduction programme converting many of the four-story maisonettes on the perimeter of the estate to attractive two-storey houses and three-storey apartments. Tenants subsequently purchased many of these. There are currently approximately 40 000 people living in Council rented accommodation on the estate.

A bus service operates between the estate and the town centre from early morning until 7PM. There are a number of shop units within a square at the centre of the estate, most of which are occupied. Each unit is protected by a heavy metal shutter which is drawn down over the complete frontage of the unit at close of business each day. Owing to repeated vandalism all telephone boxes have been removed and the only public telephones are within shops and the four public houses that serve the estate.

Brief for a local authority office

The objective of this project is to provide an area community office accommodating employees of the Council representing the departments of Housing,

Social Work and Youth and Community Work. This office will offer the opportunity to local residents to pay their rent, give notice that maintenance is required and discuss social matters.

The site for the building is on a large area of public open space on the corner of Forth Street and Tay Crescent close to Craig Square, the shopping centre.

The area is classed by the Council as 'difficult' to work in and therefore the safety of the employees and their cars and other property should be considered at all times. The office, however, should be appealing and inviting and be secure without resort to heavy metal shutters.

Accommodation

The accommodation required is as follows:

(1) Reception and rent pay desk.
(2) Two interview rooms.
(3) Waiting area including an area for children.
(4) Public toilets.
(5) Office for a supervisor and four maintenance engineers.
(6) Office for six Housing Department staff.
(7) Muster room for five rent collectors.
(8) Office for two social workers.
(9) Office for a youth and community worker.
(10) Area manager's office.
(11) Staff common room.
(12) Staff toilets.
(13) Secure storage for maintenance materials.
(14) Strongroom to which cash can be transferred on a regular basis.

ARCHITECT'S PROPOSAL

Council Area Community Office Project

Introduction

The proposed scheme illustrated in the attached drawings is considered to be the best solution to the Council's objective to construct a safe but pleasant office within the Muir housing estate. The office is designed to be constructed on four acres of public open space on the corner of Forth Street and Tay Crescent near to Craig Square, the shopping complex. The building is designed to act as an interface between the Council and the community, offering all facilities while blending in with the surroundings in a particularly novel manner.

Brief and Design Report

(1) Reception, rent pay desk and waiting area. This area should be light, airy and interesting. Part of the sitting area has a window to the garden while the other part is a full two storeys high, lit by roof lights. Furnishing in this area should be bright. Acoustics should not be too dull but permit a background buzz.

(2) Two Interview rooms each 7 m^2. The interview rooms are small, and soundproof but should be overseen from the sitting area and from reception. It is anticipated that a video link to the area manager's office will enable proceedings to be recorded. Each interview room should have a staff 'escape' door. The furniture should be so arranged as to impede the interviewee.

(3) Children's play area. An area adjacent to the sitting area of approximately 50 m^2 with toys, etc.

(4) Public toilets accessed from the garden area.

(5) Office for supervisor and four maintenance engineers – 35 m^2

(6) Office for six Housing Department staff – 80 m^2.

(7) Muster room for five rent collectors – 50 m^2.

(8) Office for two social workers – 27 m^2.

(9) Office for youth and community worker – 27 m^2.

(10) Area manager's office – 32 m^2.

(11) Staff common room – 35 m^2.

(12) Staff toilets situated centrally but inaccessible to the general public.

(13) Car parking and secure storage for maintenance materials – 700 m^2. This area on the ground floor is to be subdivided by the Council when the specific needs of car parking and storage are known.

(14) Strong room to which cash can be transferred on a regular basis by a drop operating from the area manager's office and the rent payment counter.

(15) External areas and landscaping. The location of the building is not conducive to landscaped public open space. However a feeling of open space and gardens can be incorporated within the structure.

Continues

Contd

Orientation

The north wall of the building faces out over level open ground to Stirling Tower some 500 metres away. It is proposed that this wall be the subject of a large mural, perhaps a football crowd, behind goals posts erected in front of the building. To the west is busy Forth Street. The wall facing Forth Street can also be the subject of a mural, perhaps of a more traditional building with painted windows. Tay Crescent to the south of the building provides access at ground floor level. The ground rises slowly to the east and it is proposed that with a modest amount of landscaping the three-storey building can have the appearance of a two-storey building.

Security

Security is of prime importance and for this reason it is proposed that the car parking and maintenance store be at ground floor level contained within the building. The main public areas of the office are at first floor level accessed by a swing bridge that can be swung back when the office is closed. The ground floor is accessed through secure double doors into the garage/store and through a single door to the office. The first floor garden area can be effectively secured with electronic sensors coupled to lights and an alarm. First floor reception areas and the area manager's office have a 'drop safe' facility to the strongroom below.

Summary

This rather unusual solution will be popular with local residents as, unlike neighbouring shops, the building is appealing and inviting with a total absence of external steel.

ELEMENTAL COST PLAN

District Council Area Community Office

Car park and store	736 m^2
Office	705 m^2
Total floor area	1441 m^2

	Total cost (£)	Cost per m^2
Substructure	64 975	45.09
Frame	16 528	11.47
Upper floors	82 353	57.15
Roof	22 047	15.30
Stairs	3 992	2.77
External walls	139 057	96.50
Windows and external doors	31 097	21.58
Internal walls and partitions	22 206	15.41
Internal doors	21 197	14.71
Wall finishes	13 834	9.60
Floor finishes	3 588	2.49
Ceiling finishes	7 364	5.11
Sanitary appliances	7 724	5.36
Disposal installations	1 369	0.95
Water installations	Inc	
Heat source	Inc	
Space heating and air treatment	27 494	19.08
Electrical installations	25 635	17.79
Communications installations	5 591	3.88
Builder's work in connection	3 992	2.77
Site works	27 739	19.25
Bridge, M&E and BWIC	82 451	57.22
Drainage	11 052	7.67
External services	15 174	10.53
Subtotal	636 458	441.68
Price and design risk	95 469	66.25
TOTAL	731 927	507.93

SPECIFICATION NOTES

District Council Area Community Office

Substructure

Reinforced concrete strip foundation to walls, pads to columns, 300 mm ground floor slab with granolithic finish.

Superstructure

Frame
Reinforced in situ columns with splayed heads.

Roof
Timber dormer truss, Marley Mendip smooth finish concrete interlocking tiles on battens and felt. UPVC gutters and down pipes. Dormer, timber frame, 75 mm solid foam insulation, plain tiles in vertical tile hanging to dormer face, asphalt flat roof covering.

Stairs
Reinforced in situ concrete stairs, 75 mm screed to receive carpet, steel handrails with plastic cap.

External walls
400 mm reinforced concrete, 75 mm cavity with 25 mm Jablite insulation, 100 mm Thermalite blocks perimeter wall. All other walls common brick outer skin, 75 mm cavity with 25 mm Jablite insulation, 100 mm Thermalite block inner skin, 18 mm spar-dashed render.

Windows and external doors
Standard aluminium double-glazed pivot windows in hardwood surrounds. Hardwood solid doors.

Internal walls and partitions
100 mm blockwork walls first floor, stud partition second floor. Strong room 500 mm reinforced concrete walls. Provisional sum £5000 for safe.

Internal doors
Half-hour fire resisting flush veneered doors.

Continues

Contd

Wall finishes
First floor plaster with emulsion. Second floor dry lining with emulsion. PC £8000 for murals.

Floor finishes
Fitted carpet.

Ceiling finish
Suspended ceiling with acoustic tile. Translucent plastic panels in interview rooms.

Services
Low pressure hot water from gas boiler on ground floor. Radiators, electrical, lights, power sockets, wiring boiler, alarms, hand dryers, etc.

External
Planting feature, paving slabs on asphalt, 25 m access road, 100 m footpath, soil grading.

Case Study 323

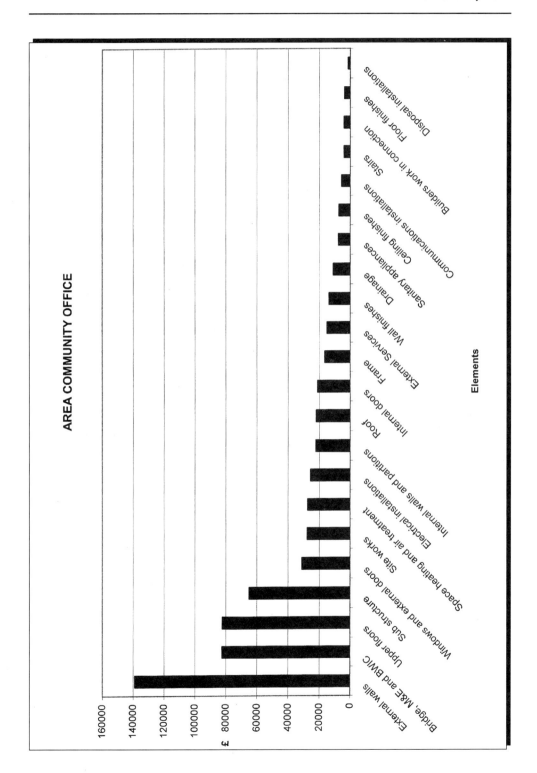

324 Appendix 2

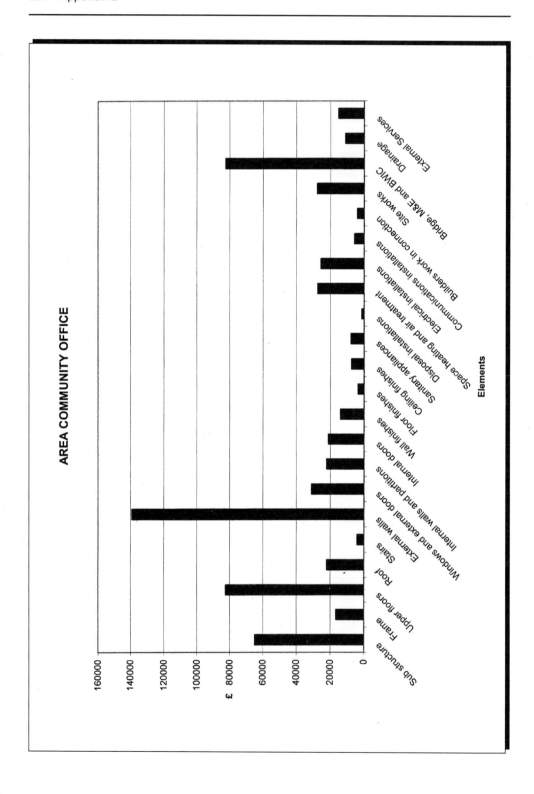

Case Study 325

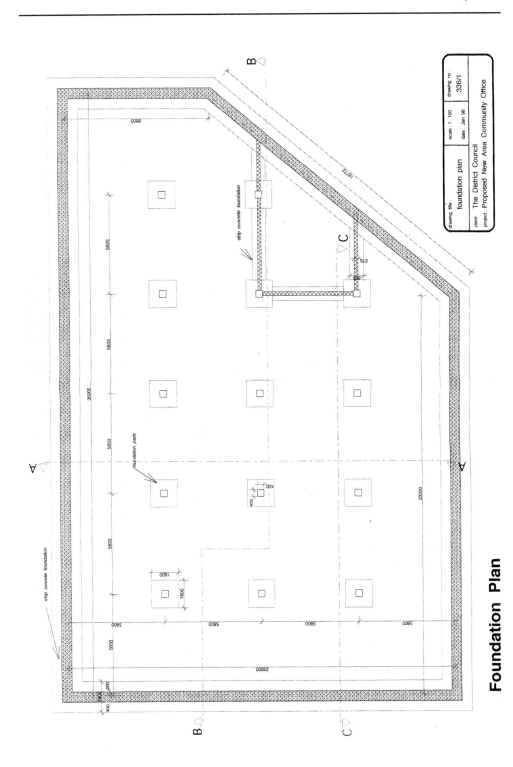

Foundation Plan

326 Appendix 2

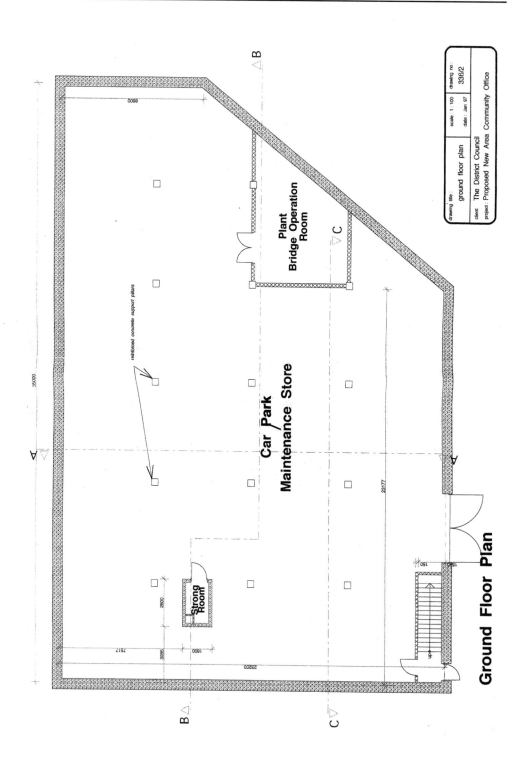

Case Study 327

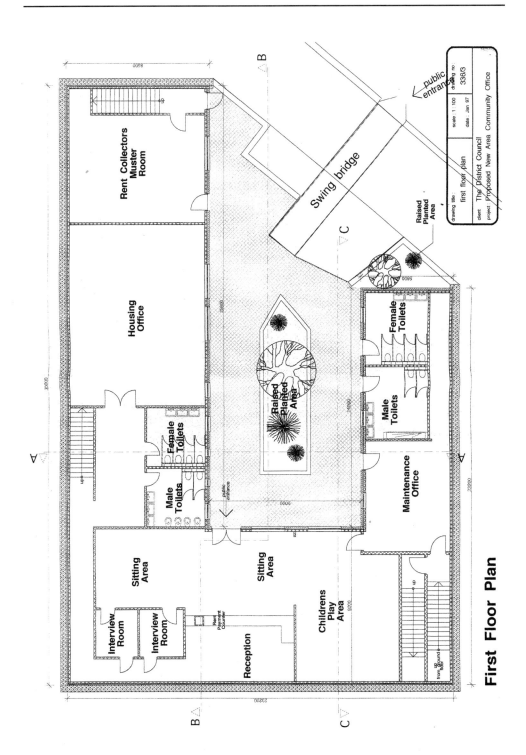

328 Appendix 2

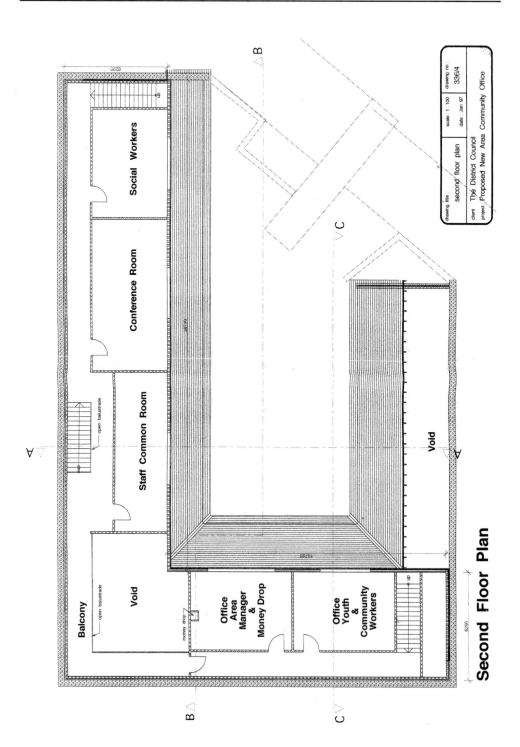

Case Study **329**

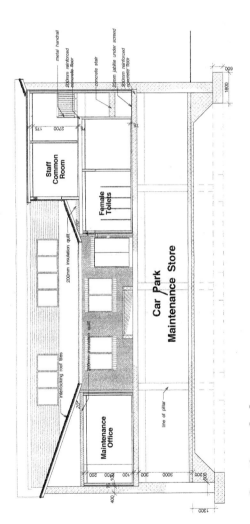

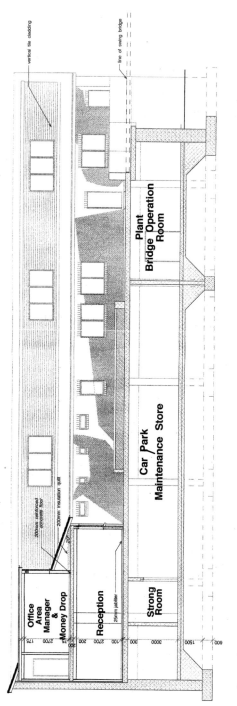

Case Study 331

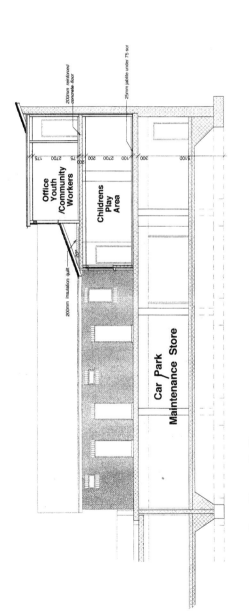

Appendix 2

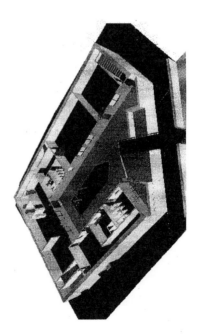

Prospectives

VM study at elemental level

A value management study at elemental level is the most appropriate given the information available. The essence of this type of study is to ascertain the function of the elements and to determine whether those functions may be provided in a more cost effective manner without prejudice to time or quality. The following demonstrates a study that concentrates on element function.

Element function stage 1: information

The first action to be undertaken as part of the information stage in element function analysis is to analyse and prepare a histogram of the elemental costs as above. In comparing these elemental costs to benchmark projects, i.e. those projects of similar nature for which element costs have previously been obtained, it is possible to determine which elemental costs are significantly higher than their benchmark project counterparts. In the example case study these elements are:

- Frame – the benchmark project does not incorporate a frame.
- Upper floors – the upper floors are significantly more expensive than those of the benchmark project.
- External walls – the external walls are significantly more expensive than those of the benchmark project.

Having highlighted the above elements as being those that appear at first sight to offer poor value for money the functions of those elements are determined.

The functions of the frame are:

(1) To transfer load.
(2) To support upper floors.
(3) To support roof.
(4) To facilitate clear spans.

The functions of the upper floors are:

(1) To support load.
(2) To distribute load.
(3) To provide a level surface.
(4) To support finishings.
(5) To accommodate services.

The functions of the external walls are:

(1) To contain space.
(2) To repel weather.
(3) To minimise heat loss.
(4) To minimise noises intrusion.
(5) To resist vandalism.
(6) To resist damage.
(7) To support floors.
(8) To support roof.
(9) To facilitate openings, for example of windows, doors, etc.
(10) To aesthetically enhance structure.
(11) To support fittings, fixtures, etc.

Element function stage 2: brainstorming, evaluation and development

The element functions are considered in turn by the team and brainstorming alternatives are noted. An advantages/disadvantages exercise is undertaken following brainstorming (see Table A2.1).

Table A2.1 Advantages/disadvantages exercise.

	Idea	Advantages	Disadvantages
1	Pre-cast in lieu of in situ concrete frame	Shorter construction on site Quality assured	Expensive?
2	Steel frame in lieu of in situ concrete frame	As above	None
3	Masonry support in lieu of frame	Does not introduce new trade on site	Longer spans are difficult
4	Split level building up with semi-basement, upper floor largely on ground	Reduces the area of upper floor	The need to find a sloping site
5	Masonry walls in lieu of in situ concrete walls	Does not introduce new trade on site	Longer spans are difficult
6	Pre-cast cladding in lieu of in situ concrete walls	Makes good use of frame Reduces foundations Opportunity for aesthetic features	None
7	Lightweight structure surrounded with thorn bushes	Less expensive	Deterrence reliant upon the effectiveness of thorn bushes
8	Reinforced earth embankment planted with thorn bushes to first floor level	Less expensive? Aesthetically pleasing	As above

Following the advantages and disadvantages analysis, the team decided to take two ideas for further development, these being:

- A steel frame, permanent formwork and in situ concrete deck with pre-cast concrete panels to the perimeter. The team also decided to investigate the incorporation of ventilation louvres in the pre-cast concrete panels to allow natural ventilation of the car park area.
- A totally masonry solution incorporating load bearing walls supporting pre-stressed pre-cast concrete planks. The external walls would be constructed of two skins of rendered block work with air vents to allow natural ventilation of the car park area.

Box A1.2   Debate

The VM study at elemental level is valid value engineering to determine the most cost effective solution to the functional requirements of the elements. However, the search for value improvement in this instance is more effectively conducted through the determination of those functions provided by all three elements together. The team decided that the primary functions provided by the three elements were to protect staff cars, to secure money and to secure materials. The question to be asked is should the client spend almost a quarter of a million pounds on these three functions? Or to put it another way, does spending a quarter of a million pounds on these three functions offer good value for money? The team decided that the answer to the latter question was no and began brainstorming solutions to these three functions.

Element function – holistic approach stage 1: information

As outlined in the debate above it may be necessary to identify the primary functions of a number of elements that appear not to be providing value for money. In the case study example it can be seen that a large sum of money is being spent on three functions namely:

- Protect staff cars
- Secure money
- Secure materials

Brainstorming at this level is only worth undertaking if it is accepted by the client and the rest of the team that significant changes to the geometric shape of the building may occur. It is important to stress this as in some circumstances a re-application for planning permission caused by a change in shape may not be advantageous to the client.

Element function stage 2: brainstorming, evaluation and development

The functions – protect staff cars, secure money and secure materials – are considered in turn by the team and brainstorming alternatives are noted. An advantages/disadvantages exercise is undertaken following brainstorming (Table A2.2).

Table A2.2 Advantages/disadvantages exercise.

	Idea	Advantages	Disadvantages
1	External car park with perimeter fencing	Relatively inexpensive to construct, saves enclosed space	Vulnerable to missiles thrown over the fence
2	Ditto with CCTV	As above but with the deterrence offered by CCTV	Continuous maintenance
3	Configure buildings as a single or two-storey structure with a central courtyard for parking	As secure as a basement with lower cost	Larger building footprint and the loss of the public garden
4	Ditto but with double courtyard, one for public with garden and one for staff parking	As above, gives public facility as existing	Even larger building footprint
5	Material storage remote from community office as at present	Removes requirement for secure material storage.	May not meet the primary requirement for a fast response
6	Negotiate with builders merchants for just-in-time deliveries	As above	May compromise competitive purchasing
7	Parking of maintenance vans overnight remote from community office, i.e. as at present	Removes secure overnight parking of vans	None
8	Parking of rent collectors' vans overnight remote from the community office, i.e. as at present	As above	Determines where rent collectors commence work

Following the advantages and disadvantages analysis, the team decide to take two ideas for further development, these being:

- A single-storey or two-storey building with courtyard or external fenced parking and CCTV.
- The investigation of just-in-time deliveries by builder's merchants.

> **Box A1.2** Debate
>
> The holistic approach to elemental value engineering is attractive from the perspective of logic and team acceptance. However, because its logic is based upon a limited backwards investigation it suffers from the lost opportunity of going right back to the beginning to the point of being perfectly clear of the client's objectives and the client's value system.
>
> The example that follows is an illustration of going right back to the beginning and using the tools and techniques described earlier to discover the true facts.

An illustration of full value management

Stage 1: pre-workshop information

From the pre-brief workshop information phase techniques described in the text the following techniques were selected by the facilitator as being the most appropriate for the case study project:

- Interviews.
- Document analysis.
- Walkthrough of a similar facility.
- Issues analysis.

Of the above techniques it was determined, through discussion with the director of housing (the client for this project) that document analysis would yield little in terms of the background to the project and that interviews with key personnel would be the most beneficial. It was also realised that as this was the first of a proposed series of area office projects a walkthrough of a similar facility would not be possible.

Five interviews were arranged with the personnel most closely associated with the project. This gave the facilitator an opportunity not only to gather information but also to inform these influential client personnel of the aims and objectives of the workshop and to outline the basic principles of value management.

INTERVIEW RECORD 1

**Taped interview with Harry Jones,
Director of Social Work and Welfares Services**

Harry Jones is Director of Social Work and Welfare Services and is responsible directly to the Chief Executive for:

- Administration department at central office.
- One adult training and resource centre.
- One children's centre.
- Two homes and four day centres for older people.

Mr Jones stated:

'The proposed Area Community Office is an interesting project from my perspective as it is being organised and funded by the Housing Department but has a major social work perspective. Don't get me wrong, I believe that the idea of taking staff into the war zone (don't write that down) is a good idea but the fact is that this is where my staff already work on a daily basis.

This particular area, the Muir housing scheme, is very deprived and is on a downward spiral. We have a very specific project here which is to dedicate two members of staff with the specific aim of getting started a cooperative children's play school. One major problem in Muir is the large number of single parent families, most of them young women with one or two children. The reason that they end up on benefit in a grim flat in Muir is that they have left their parental home or been thrown out. Our plan is to select and train a small number of men and women to run a play school and child care centre. Once this is commenced it will allow many of the single parents to leave their children in a safe environment and find work. The cooperative element comes from the fact that we have to encourage those who work to pay for the child care services provided by others. These child carers will probably have to work strange hours to cover the part time and shift work that is a feature of the employment in this area.

I have obtained funding from the council to cover the cost of two social workers for four years and a small amount which will cover start up costs of the play school and child care centre. We hope to house this on the local primary school site – not that they have any spare classrooms but we hope to get some temporary classrooms brought in. I can start my four years' funding any time during the current financial year but I have to say I am keen to get started within the next two to three months. The two volunteers have already been selected and appear very keen.

So there you have it. Over a four year period two social workers will be based eventually at the Area Community Office to get this project up and running.'

> **INTERVIEW RECORD 2**
>
> **Taped interview with David MacDonald**
> **Director of Leisure Services**
>
> David MacDonald is Director of Leisure Services and is responsible directly to the Chief Executive for:
>
> Administration Department at central office.
> Three sports centres.
> Various club facilities leased from the local authority – rugby club, bowls club, etc.
>
> Mr MacDonald stated:
>
> 'You probably already know that the proposed Area Community Office is being organised and funded by the Housing Department. I believe that Social Work have some input as well.
> I've an unusual job here in that I have to get started some youth activities. I don't look forward to this project and I have said that it should fall within the remit of either Education or Social Work. However, at least I don't have to go down there, not yet anyway.
> My remit is to appoint a new youth and community worker, which I have done. He has been posted off to Muir and now has the responsibility of getting youth activities under way. It is planned that he will work from the new office. He seems a keen chap and has already made contact with the local church communities. The strange thing is I've only got funding for three years and he's on a three year fixed term contract. I don't know when the building will be finished.'

> **INTERVIEW RECORD 3**
>
> **Taped interview with Roger Cowley**
> **Architect**
>
> Roger Cowley is a senior project architect with Wolsey Ford & Partners, a small practice of architects based in the City. This is the first job that Wolsey Ford have undertaken for the Council that was won in a competitive fee bid.
>
> Mr Cowley stated:
>
> 'Thank you for explaining value management to me. I have not come across the technique before and prior to your visit was somewhat apprehensive. The design evolved to date is very much based on the client's brief and a very detailed discussion with Mr Hill. This included a one-day visit to see the site during which we had the opportunity to speak to a number of shopkeepers at Craig Square. This highlighted the problems caused by vandalism. It is very difficult to conceive of a design that is inviting and welcoming and yet is as hard as nails. Some people have already compared my design to a fort and I guess that this will become its nickname – I can live with that.
>
> 'In the design of the building I had some conversations with a colleague of mine who is a structural engineer. I should emphasise that to date no quantity surveyor or structural engineer has been appointed although the Technical Services Department of the Council has prepared the cost estimate. I originally understood that the budget of half a million pounds excluded fees, VAT, fixtures and fittings, so you can imagine my surprise when it was suddenly announced that it did not.
>
> 'I will be interested to see the value management exercise in practice although I am still apprehensive with regard to any redesign work, since we took this job on a very tight fee to get in with the Council.'

INTERVIEW RECORD 4

Taped interview with Angus Hill
Director of Housing

Angus Hill is Director of Housing and is responsible directly to the Chief Executive for:

The administration of Housing, currently at the city centre office, including:
House management
House maintenance
Rent collection
Responsibility for the homeless

Mr Hill stated:

'This is a major project for me and the outcome of years of lobbying. When I joined the Council 15 years ago I said "housing has to be administered from within the Housing", otherwise the staff have no feel for what's going on. Now 15 years later we begin the first project. I've a building budget of £0.5 million which has to include all costs, fees, VAT, etc.

'I am placing three separate sections there, and there will be representation of 'Social Work and a youth worker. My sections are:

- *Administration:* This will deal with all housing queries including the homeless, bad neighbour problems, rent collection, maintenance, etc. Our aim is immediate response. Someone comes in with a problem; we'll have it dealt with within minutes instead of within hours or even days as at present.
- *Maintenance:* There will be three guys who will fix anything. We are not talking programme maintenance here, not the painting every five or six years, but the leaking hot water cylinder, the leaking roof, the broken window, even – if time – the prompt removal of graffiti. The maintenance team of three will be permanently on site and will respond quickly, having a store of materials to hand.
- *Rent collection:* Most rent is paid by the Benefits Agency direct but for those who pay, often in cash, there must be a payment counter locally, not here in the centre of the city. For those who don't pay, the rent collector will call. Each day the rent collectors will be out on the estate collecting or making arrangements to collect on a particular day at a particular time. They get their instructions each morning from head office, fed through to the Area Community Office. Then they are off, returning to the Area Office for lunch and to deposit their takings. Then off again. They have distinctive vans that will be stored at the area office.

'So there it is. In two years I'll be able to realise a dream. Then who knows, in two years the Chief Executive of the Council retires so it's in my interest to make a good job of it here. Certainly if it all works out this will definitely be the first of a series of similar projects.'

INTERVIEW RECORD 5

Taped interview with Alan Hooper
Maintenance Superintendent (Muir Estate)

Alan Hooper is responsible to the Director of Housing for:

Responsive maintenance and repairs on the Muir estate.
Supervision of two joiners, a plumber and an electrician.
Reporting on planned maintenance items.
Inspecting planned maintenance work undertaken by contractors.

Mr Hooper stated:

'I'm all in favour of the planned Community Office as it will make my work very much more simple. I live on the estate, I've lived on the estate since it was built. At present I catch a bus, 20 minutes' ride to the depot to collect a van to drive back virtually to where I live. The Community Office means I can walk to work. One of the joiners and the electrician also live on the estate. We've all bought our own homes. Clocking on for them is going to be easier as well. Of course we can't bring the vans home and park them in the estate as they would be demolished overnight.

'Most of the work we carry out is what you'd expect in a 1960s estate – repairing corroded plumbing, tending to electrical faults usually caused by damp, replacing rotten doors and an occasional casement.

'We tend to get left alone, you don't see many Council officials down here. Most of the people on the estate are OK, it's the youngsters who cause most of the damage. Many of them are not from around here.

'I notice we are going to get a new store. I'm not sure why that is as we have only just reached an agreement with the local builder's merchant to effectively act as our store. Since the builder's merchant is only three-quarters of a mile away this is a much better arrangement. The lads can sign for anything provided that the job voucher can be correlated with the invoice. It is a simple system really:

(1) The tenant phones saying "I've got a leaking pipe".
(2) I'm phoned on my mobile and go around there and inspect the job
(3) I raise the job voucher and give it to the lads.
(4) The lads record their time and materials, which they get from the builder's merchant.
(5) I attach the materials received summary from the builder's merchant and sign the voucher.
(6) All the vouchers get returned to headquarters who correlate them with the builder's merchant's invoice.

In reality, because there are no phones, the tenants catch me in the street and use my mobile to phone headquarters. It's interesting to see how long it takes them to phone me back, usually the job's half done by then. Me and the lads keep the estate in pretty good nick, and there's not usually too much complaint about overspending, which we do every year.'

Stage 2: pre-workshop team, agenda and location

Subsequent to the completion of the interviews and in discussions with Mr Hill, the membership of the team was determined using the ACID test. The agenda for the workshop was determined by reference to the standard form for a pre-brief and briefing study.

The workshop team comprised:

Mr Hill, Director of Housing and the project's sponsor
Mr MacDonald, Director of Leisure Services
Mr Jones, Director of Social Work and Welfare Services
Mr Cowley, architect, Wolsey Ford & Partners
Mr Smith, quantity surveyor, Technical Services Department
Mr Hunter, structural engineer, Technical Services Department
Mr Reid, project manager designate, Technical Services Department
Reverend Cunningham, local church minister
Mr Fraser, local councillor
Mrs Jones, clerk, House Management Department
Mr Hooper, maintenance superintendent
Mrs MacLean, secretary, Muir Residents Association
Mr Mackenzie, rent collector (final day only)

In addition there were two facilitators and a recorder.

The venue for the workshop, arranged through the offices of Mr MacDonald, was a residential water activity centre, located in a neighbouring town. The centre had a large, well-equipped training room that was ideal for the workshop.

The agenda for the two-and-a-half day workshop was as shown in Fig. A2.1 (see page 344).

Stage 3: workshop

Issues analysis

The various techniques for exploring the issues are described in Chapter 3. For this project the facilitator decided that the group contained the key stakeholders to the project who were therefore well acquainted with the issues. The facilitator directed the issues analysis as a brainstorming session of all topics, recording onto repositional sticky notes and taking any subject in random order. The team worked from their own knowledge and interpretation of the facts and called out items that they considered as being important without discussion or debate. This exercise lasted approximately 45 minutes. The sticky notes were randomly distributed on a large sheet of paper.

The sticky notes were subsequently organised by the team under headings, for example 'client organisation', 'users', etc. The headings were taken from the standard list and added to as appropriate. The result of this exercise was a working wall containing the notes organised under headings.

Time	Activity
Day 1	
15.00	Meet at the training centre, check in and take minibus to the Muir estate for a briefing by Mr Hill and Mr Jones. Inspection of the site
18.00	Return to training centre for an overview of value management and an introduction to the agenda
19.00	Dinner and 'team building'
Day 2	
09.00	Introduction to the day's agenda
09.15	Issues analysis
10.15	Coffee
10.30	Client value system
11.00	Identify important issues
11.15	Timeline
11.30	Expand on issues
11.45	Brainstorm strategic functions
12.15	Initial sort of functions – highest ranking need to lowest ranking wants
12.45	Commence construction of function diagram – agree mission of project
13.00	Lunch
14.00	Complete construction of function diagram
14.45	Review information and function diagram
15.00	Tea
15.15	REDReSS analysis
15.45	Agree strategic brief as represented by function diagram and supplementary information

Option 1: to brainstorm solutions to the strategic brief and complete an action plan.

Time	Activity
16.00	Brainstorm solutions to strategic functions prior to option appraisal
17.00	Select options to be carried forward to option appraisal
17.30	Complete action plan
18.00	End of workshop

Option 2: to continue the workshop to complete the foundation for the strategic brief and project brief.

Time	Activity
16.15	Identify users and user flow analysis
17.45	Identify spaces from user flow analysis
18.15	End of Day 2
19.15	Dinner
Day 3	
09.00	Information review
09.30	Functional space analysis
10.30	Coffee
10.45	Adjacency matrix
11.15	Highlight functions for brainstorming
11.45	Brainstorming
12.45	Initial sort of ideas
13.00	Lunch
14.00	Select and group ideas for development
14.30	Outline development in groups
15.00	Tea
15.15	Outline development in groups continued
16.30	Presentations by groups
17.00	Action plan
18.00	Sum up and end

Fig. A2.1 Brainstormed functions.

Client value system

The client value system is used to obtain ordinal measurement in the form of a ranking of the client's values. The client value system demonstrated that the three most important values were, in order, politics/community and operating cost jointly valued with comfort. This indicated that the built facility must take into account politics and the local community by satisfying their needs, it must consider the life cycle costs of the facility and the design focus must be internal rather than external and directed towards the comfortable and efficient use of the customers' space.

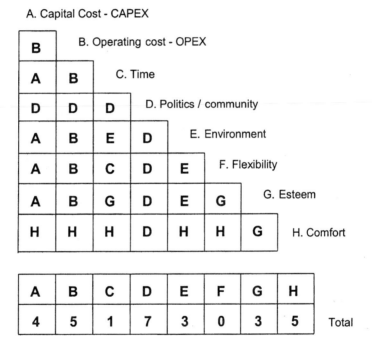

Fig. A2.2 Priority matrix.

Determine importance from the issues analysis

After establishing the client value system each team member was given eight black dots to 'spend' on those items judged to be of importance. Once the black dots were spent each team member was given a further three red dots to spend on those items which they considered were so important that unless the topic were fully understood or resolved the project might fail. This is, from the individual's perception, a fairly safe activity since the spending of dots happens relatively quickly with everyone contributing at the same time. It would be unusual for a team member to stand back to observe where colleagues were placing their dots. The activity is therefore anonymous. It is only when the team stand back that the graphic effect of the working wall is appreciated. The complete issues analysis is given below.

Project drivers
Community need	2B	
National precedent	2B	
Quality of construction	1B	
Partnering ethos	1B	
Accuracy of brief	5B	4R
Local presence		
Availability of funds	4B	1R
Collection of rentals		
Personal egos		
Community/Council interaction		
City Hall enforced policy		

Change management
Project financial controls	4B	1R
Notification system		
Risk analysis	1B	

Security
Money security		
Van (maintenance) security		
Vandalism of finished product		
Segregation of public and staff		
Attracting unwanted outsiders	1B	
Safe parking	1B	
Night security		
Secure cycle parking		
Employee safety	3B	4R

Community
Resentment of locals to waste of money	1B	
Staff/public interface		
Local employment need		
Access to town centre		
Local community issues	1B	1R
Resident consultation	1B	2R
Double glazing for local tenants	1B	
Bigger play area for children		
Local community vetos		
Resident training and education		
Good PR		
Jobs for locals	1B	

Disruption to community   1B
Impact on community   2B
Enhancing education facilities
After hours use
Robust public areas
Space for youth activities
Access to public phones
Impact on existing infrastructure
Drugs problem
Ratio of public : private areas
Future nursery accommodation

Central/local politics
Changes to local government
Local authority budgets in future
Community share in savings
Planning issues
Consultation at all stages
Involvement with design by Council

Finance
£8000 mural
Annuality
Community worker 2 years' funding
Project funding   3B   3R
Life cycle funding   1B
Future funding for community worker   1B
Whole life cost   1B
Ultimate disposal costs
Project cost outside budget
EU/regeneration grants   1B
Taking money out of local economy   1B
Additional funding
Collect rents
Inflationary pressure on budget
Consultants' fees
Reducing cost   1B

Time
How soon facility is available   6B   3R
Speed of construction   2B
Opening date   1B
Programme overlaps
Budget and programme overruns   4B   1R

Appendix 2

Legal/contractual

Supervision of build (CoW)	1B	1R
Decision to proceed	2B	3R
Insurance issues		
Dispute resolution	1B	
Form of contract	2B	
Level of LADs		

Project design

Contractor's innovation	4B	2R
Maintainable landscaping		
Access for elderly		
Part L Building Regulations		
Internal finishes speculations		
Window cleaning regime		
Future role of concept architect		
Accurate soils information		
Proper drying and changing facilities	1B	
Staff/public interface	1B	
Extent of design changes funding	3B	
Noise levels		
Maintenance regime	1B	1R
Aesthetic appeal	1B	
Planned maintenance life cycles		
Fear of bridge		
Buildability		
Disabled access		
Operational functionality	1B	
Quality/cost decision making	1B	
Vandalism/security	4B	
CABE issues		
Materials storage		
Lifetime of building		
Maintenance office		
Proper access to roof for maintenance		
Swing bridge maintenance safety	1B	
Shower area		
Created working environment		
Welcoming environment staff/users		

Organisation

Operational functionality	6B	6R
Big brother effect	1B	
Users' attitudes		
Popularity with users		

Integration of departments' employees
Increased maintenance staff
Department communications
Staff communications   1B
Space planning
Staff retention

Context
Defined users coming into building
Contact with other support services, police, fire
Green issues
Sustainable development
Award winning
Space planning
Internal environment

Location
Site location in estate
Construction traffic

Timeline

The group identified where the critical points in the project were on a timeline from the present date until 2007, where the last critical point in the project was expected to lie (Fig. A2.3). The timeline was roughly drawn on a flip chart with the aim of sensitising the team to the critical anticipated dates.

Project drivers

In this particular workshop the team called out project drivers with the issues analysis. If this does not happen the facilitator should elicit from the team at this stage those items that are driving the project. This may be done under headings as illustrated in Table A2.3.

Expand on issues

At this stage the working wall comprises the dotted issues analysis, the client's value system, the timeline and project drivers. The facilitator at this point returned to the working wall and specifically to those items in the issues analysis designated as important by the number of dots. The facilitator asked for further comment on the heavily dotted issues. This stage of the workshop can become tense as frank and honest comment is elicited on topics that may be politically sensitive or of primary importance to a member of the team. The statements that follow are those that resulted from an expansion of those issues indicated as being significant.

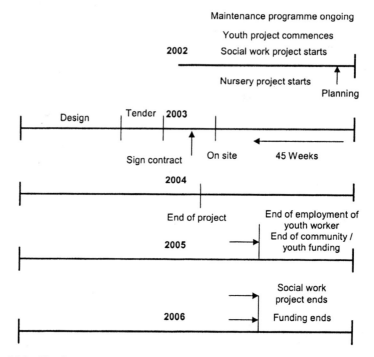

Fig. A2.3 Timeline.

The team understood that there was not total commitment from all departments or indeed from the local authority itself. It was recognised that commitment might disappear in the event of staff being moved around, and it was seen to be the first scheme of its type. The project might need to be viewed as a part of local services, shops, medical centre, etc. It needed to be open access, with particular attention to access for the disabled, and it might be necessary to encourage the diversion of some bus routes to pass the facility. Evening availability of services was seen to be paramount. The team questioned whether the current design would attract people and whether there was sufficient access for security vehicles into the basement car park. Finally the team queried whether rent collection in the centre of the estate was necessary.

It was recognised that the housing function is the only permanent function ascribed to the office, the community worker function and the social work project are both supported through fixed term funding. A query was raised on whether these two activities should be based at the office.

Staff and money security was seen as being a major issue. Encouraging complaints might also endanger staff. This is a small facility and therefore staff were seen as being vulnerable. Finally the security of cars had been emphasised but staff might not arrive in cars.

Was the project seen as an advantage locally and could it succeed against a background of antisocial behaviour? The project is relying heavily on local

Table A2.3

Time drivers	Cost drivers
2 years to realise a dream 2 social workers for 4 years 1 youth worker for 3 years Waiting 15 years for the project Building budget £0.5 m now CEO retires in 2 years	Brief Location Security requirements Enhancing neighbourhood Client's operational organisation Accessibility while maintaining security
Quality drivers	Project drivers
Budget Security Enhancing environment Staff expectations	Angus Hill Community policy Efficiency gains Desire to upgrade estate Budget availability
Space drivers	Design drivers
Occupancy Car parking Security Storage	Security Aspirations to enhancing environment Space requirement Location Organisational arrangements Safety Variety of functional uses Robustness of facility

cooperation, particularly in the area of child care. The team was concerned that the end users' view of the project was not clear and there appeared to be a need for communication/marketing to the local population and to ensure a sense of local ownership.

The local authority departments appeared to be discrete organisations with a lack of interaction. There seemed to be a need to break down barriers between departments both in terms of central agencies such as the Benefits Agency and locally. The Housing Department had the whole budget for the building but this budget might be insufficient for the investment required of it. The existing brief did recognise that if all parties were committed to the project there could be savings particularly with the involvement of local people. If successful this project might be repeated in other areas. The building was to be seen as a place for resolving all problems and therefore the team believed that there was a mismatch between a quick response and a sitting area and children's play area. There was also a belief that the circulation areas were excessive.

Half-a-million pounds had been budgeted for the project but based on the initial design and cost plan the project was over budget on construction cost alone. There seemed to be a fixed budget for the social and youth projects that was time related. It was noted that the cost planned figure for the building was without fittings and fixtures, fees and VAT. The team queried the exploration of external funding. There had been no consideration of financial risk. Running costs were seen to be high and this would be repeated in the event of replicated projects in other areas. Cost of the project, however might be offset by reduced operating costs of the estate.

The project was effectively an experiment and therefore could be viewed as having a short life. The life of the project therefore needed to be clearly established as it would influence the maintenance strategy and perhaps restrict the standards or definition of finishes.

REDReSS

The REDReSS exercise can be undertaken at any point following the issues analysis to correct any misconceptions which might be creeping into team thinking or to highlight any factors the significance of which has not fully been taken on board. The REDReSS on the project brought out the following points.

Re-organisation
Facility flexibility

Expansion
Nursery and youth & community
Space to be sufficiently large

Disposal
Building to match estate
60-year life
Recycling a contractor issue

Refurbishment and maintenance
Tough fittings
Cheap/simple/easy
Low energy fittings
Easy glass cleaning

Safety and security
Secure by design
Light open areas
Vandal proof
Site accident free
Site to be fully secure
Staff by design
Users by design

Time, cost and quality

At this point the facilitator decided that sufficient confusion surrounded the housing, youth and community and social work projects to re-examine them using a simple time, cost and quality triangle (Fig. A2.4). In many respects this duplicates the client value system but was used in this instance to clarify the context of the overall project. This was a good example of the facilitator deviating from the agenda to resolve an issue that might otherwise have been disruptive in the background.

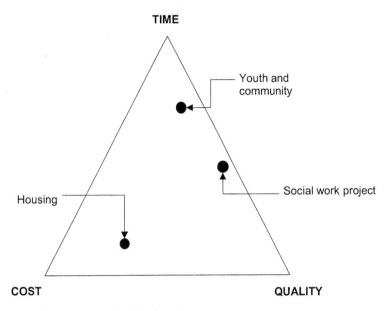

Fig. A2.4 Time, cost and quality triangle.

The time, cost and quality triangle highlighted differences in the three facets of the project. Once in the open these can be explored further through the function diagram.

Function diagram

There are three steps to the construction of a function diagram:

(1) Brainstorm functions as noun plus verb definitions.
(2) Construct project function priority matrix.
(3) Construct function diagram.

The team undertook the brainstorming and the brainstormed functions were written down on sticky notes in the random order in which they were generated. The facilitator prompted for functions by continual reference back to the working wall. The list of brainstormed functions is shown in Table A2.4.

The project function priority matrix is constructed by asking of each sticky

Table A2.4

Establish needs	Establish multi-function integration
Ensure flexibility	Create value
Meet community needs	Establish local Council presence
Support community	Secure funding
Involve community	Extend funding
Establish project brief	Control finances
Secure environment	Increase innovation
Achieve longevity	Control programme
Enhance communication	Enhance environment
Ensure staff comfort	Increase local employment
Ensure user comfort	Improve user interface
Engender community spirit	Ensure continuity
Change community perception	Improve programme
Ensure operability	Create pleasing environment

note function whether it is technical or strategic and whether it constitutes a need or a want. The note is transferred to the appropriate box in the matrix where its position is ordered relative to the other functions in the box (Table A2.5). The higher priorities are listed at the tops of the respective boxes.

A strategic or customer orientated FAST diagram is constructed by focusing on the strategic needs and wants (Fig. A2.5). The highest order needs tend to form the mission of the project with supporting functions being positioned to the right. The strategic wants tend to be positioned below the centreline of the mission statement.

Table A2.5

Strategic needs	Technical needs
Meet community needs	Establish needs
Involve community	Establish project brief
Establish local Council presence	Secure environment
Ensure continuity	Achieve longevity
Ensure operability	Ensure staff comfort
Ensure flexibility	Ensure user comfort
Establish multi-function integration	
Change community perception	
Enhance communication	
Engender community spirit	
Support community	
Improve user interface	
Change perception of community	
Strategic wants	Technical wants
Create value	Control finances
Secure funding	Increase innovation
Increase local employment	Control programme
Extend funding	Improve programme
Enhance environment	Create pleasing environment

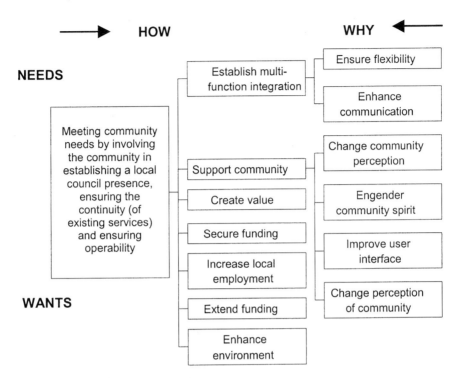

Fig. A2.5 Strategic FAST diagram.

The mission statement will require word crafting to make it read as a flowing statement. It is important to have unanimous team agreement that the statement truly reflects the mission of the project, remembering that the mission at this stage does not necessarily imply a building.

Goal and systems modelling

In order to further inform the team on the decisions to be taken as an outcome of the function analysis stage a goal and systems model can be constructed. The goal and systems model is constructed by taking the strategic needs and wants (the goals) from the extreme right side of the FAST diagram and relating these to the systems currently available for carrying out the goal activities (Fig. A2.6). This indicates how many of the systems in place answer the goals of the project. What is illustrated here is prima facie evidence that the Housing Department actually does little to meet the requirements of the community.

Workshop review

There should be a pause in the workshop at this stage to consider the requirement to build. There are three outcomes to this debate:

356 Appendix 2

(1) The client wishes to reconsider the project and the workshop closes.
(2) The client wishes to brainstorm options for an option appraisal.
(3) The decision to build is affirmed and the workshop moves to the project briefing stage.

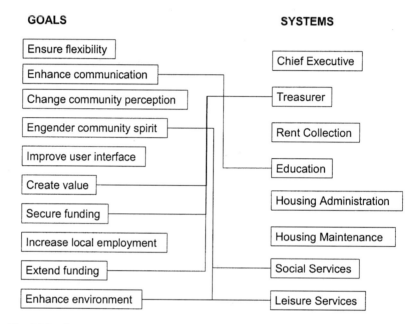

Fig. A2.6 Goal and systems model.

In this case study the decision was to build a facility and therefore the team moved forward to consider functional space. It is at this stage that the technical needs and wants left behind from the construction of the FAST diagram, which is by definition strategic, are considered.

User analysis

The user analysis is the first step in the process of deriving a specification of functional space, the foundation of the project brief. This exercise is carried out by identifying all of the users of the space. In this exercise the users were identified as general public, maintenance operatives, social workers, youth and community worker, area manager, housing administration staff and rent collectors.

For each user or group of users a flow chart is constructed (Figs A2.7 to A2.12). The flow chart is configured to represent the way in which the user will go about their work in the building.

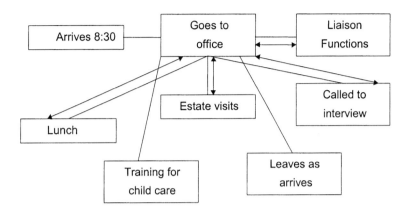

Fig. A2.7 Social worker flow chart (youth and community worker similar).

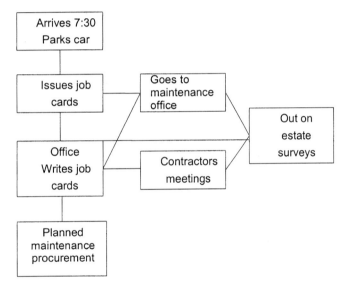

Fig. A2.8 Maintenance manager flow chart.

358   Appendix 2

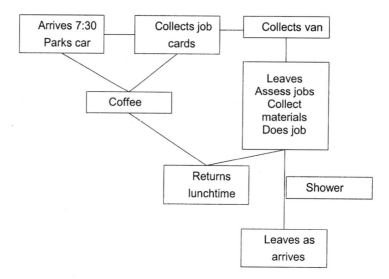

Fig. A2.9   Maintenance worker flow chart.

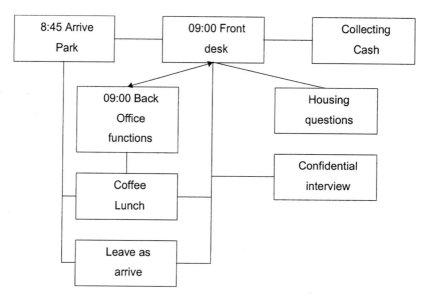

Fig. A2.10   Housing staff flow chart.

Case Study 359

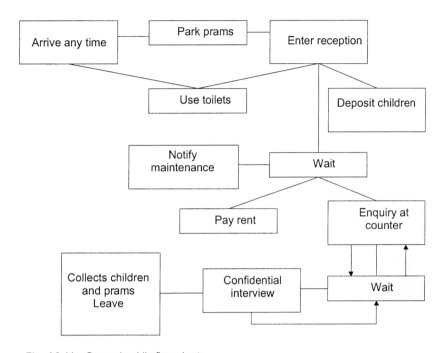

Fig. A2.11 General public flow chart.

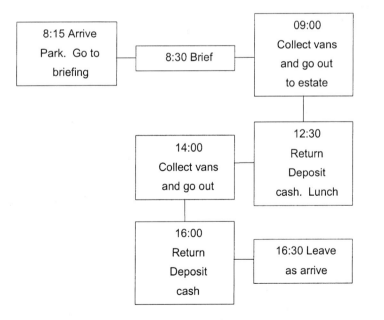

Fig. A2.12 Rent collectors flow chart.

User space analysis

Each functional activity in the user flow diagrams leads to a functional space used in a particular way. From the diagrams in Figs A2.7 to A2.12 the following user spaces were identified.

Car park – secure (9 council vans, 11 private cars)
Briefing area – rent collectors, maintenance
Cash up secure area for records
Lunch – coffee facility
Toilets/showers/changing/boot area
Maintenance desk (1)
Meeting area
Reception area
Confidential interview rooms (2)
Housing desks (4)
Storage for materials/tools
Strongroom/safe
Pram park/cycle park – Secure
Childrens play area
Toilets
Waiting area
Social worker desk (1)
Area Manager desk (1)
Youth and community desk (1)
Receptionist/filing clerk
General offices (8)
Area Manager office (1)

Adjacency matrix

The adjacency matrix (Fig. A2.13) indicates those spaces distilled from the above which need to be readily accessible (+5) or need to be isolated one from another (−5). The only data which now needs to be added to complete the outline project brief is the room specification in terms of functional size, quality, IT provision and services. This becomes the raw data from which to compile the room data sheets.

Brainstorming

The brainstorming undertaken at this stage is to give the project brief final directions with regard to the incorporation of all of the above. All those ideas that were deemed not feasible from the outset were deleted prior to a vote by giving each member of the team five sticky dots to place on those ideas which they would be willing to champion. Further analysis was conducted by decid-

										A. Car Park
0										B. Common room
+5	+2									C. Staff toilets / showers / changing / drying
-5	-2	-5								D. Public toilets
+3	0	0	-5							E. Plant room
+5	-5	-5	-5	-5						F. Money safe
0	+1	+2	-5	-3	+3					G. General office
-5	-1	-5	+5	-5	+5 / -5	+5				H. Reception / waiting / children
-5	-2	-2	+3	-5	-5	+5	+5			J. Confidential interview
+3	0	-3	-5	-5	+5	+2	-5	-5		K. Meeting room and cash up
0	0	0	-5	0	-5	+2	+2	-5	-5	L. Meeting room

Fig. A2.13 Adjacency matrix.

ing which ideas were technically feasible (TF), economically viable (EV), functionally suitable (FS) and client acceptable (CA). Some brainstormed ideas are shown in Fig. A2.14. Items that do not survive the sort process are scored out.

Action plan

The design team were tasked with completing the project brief but including large multi-use spaces that might be used as a daytime child care facility and by clubs in the evenings. The Council presence would be in the form of a help desk with limited staff dealing with all queries, social and housing. Staff would be in contact with maintenance workers, social workers and the youth and community worker via mobile phone.

		Vote	TF	EV	FS	CA
1.	Create housing help desk only at centre	13	y	y	y	y
2.	Build community centre	9	y	?	y	y
3.	Build large hall with offices attached	6	y	y	y	y
4.	Maintenance materials from builder's merchants as now	13	y	y	y	y
5.	10 lock up garages for staff cars in secure walled yard	4	y	y	y	y
6.	~~Start a local radio and use for information on e.g. maintenance~~					
7.	Install an internet server and encourage contact by e-mail	3	n	n	y	y
8.	Build a facility for social & youth project near local school	4	y	n	y	n
9.	~~Issue all tenants with 'pay as you go' mobile phones~~					
10.	~~Build temporary building for youth project~~					
11.	Build steel shelter for youth project	1				
12.	Break up projects into three distinct facilities	10	y	y	y	y
13.	Operate help desk from mobile library	1				
14.	~~Build a large concrete dome and let tenants fit out~~					
15.	Rent local hall for youth project	1				
16.	~~Start 5 a side football team~~					

Fig. A2.14 Ideas analysis.

Conclusion of case study

The case study has been presented in three different ways.

(1) As a value engineering study which preserves the design as presented and looks to element functions to drive the search for innovative solutions in order to provide the function at the required quality and the lowest cost.
(2) As a value engineering study but with the emphasis on space, recognising that a reactive seeking of innovative space solutions may result in radical changes to the design.
(3) As a value management study in which the mission or primary function of the whole project is sought and innovative solutions uncovered with the aim of providing the client with the best value for money solution.

The case study has illustrated the tools and techniques in action. With such a simple exercise it has not been necessary to edit down so what has been presented here is the entire result of a training exercise. It should be recognised, of course, that the training team was not a real team and therefore some of the comments and some of the issues would not have occurred in practice. However, it is as close to reality as can be achieved in a book of this type and the authors trust it has provided some instruction and entertainment.

Bibliography

Allport, G., Vernon, P. & Lindsey, G. (1960). *A Study of Values*, 3rd edn. Houghton Miflin, Boston, Mass.

Armstrong, J., Dixon, R. & Robinson, S. (1999) *The Decision Makers: Ethics for Engineers*. Thomas Telford, London.

Austin, S., Baldwin, A., Hammond, J., Murray, M., Root, D., Thomson, D., Thorpe, A. (2001) *Design Chains*. Thomas Telford, London.

Belbin, Meredith R. (1981) *Management Teams: Why They Succeed or Fail*. Butterworth-Heinemann, Oxford.

Belbin, Meredith R. (1993) *Team Roles at Work*. Butterworth Heinemann, Oxford

Best, R. & De Valence, G. (1999) *Building In Value: Pre-Design Issues*. Edward Arnold, London.

Bicheno, J. (2000) *The Lean Toolbox*, 2nd edn. PICSIE Books, Buckingham.

Bicheno, J. (2002) *The Quality 75: Towards Six Sigma Performance in Service and Manufacturing*. PICSIE Books, Buckingham.

Blyth, A. & Worthington, J. (2001) *Managing the Brief for Better Design*. Spon Press, London.

Borjeson, L. (1976) *Management of Project Work*. The Swedish Agency for Administrative Development, Satskontoret, Gotab, Stockholm, Sweden.

Broadbent, G. (1973) *Design in architecture*. Wiley, Chichester.

BS EN 12973:2000 *Value Management*. British Standards Institution, London.

Burt, M. E. (1975) *A Survey of Quality and Value in Building*. Building Research Establishment, Garston, Watford, Herts.

Bytheway, C. W. (1965) Basic function determination technique. *Proceedings of the Fifth National Meeting of the Society of American Value Engineers*, **2**, 21–3.

Cameron, Esther (1998) *Facilitation Made Easy*. Kogan Page, London.

Checkland, P. B. (1981) Science and the systems movement. In: *Systems Behaviour* (Open Systems Group) 3rd edn, pp. 288–314. Harper & Row, London.

Christopher, M. (1998) *Logistics and Supply Chain Management: Strategies for Reducing Cost and Improving Service*, 2nd edn. p. 10 *Financial Times* & Pitman Publishing, London.

Churns, A. B. & Bryant, D. T. (1984) Studying the client's role in construction management. *Construction Management and Economics*, **2** (1), 177–84.

Crum, L. W. (1971) *Value Engineering: the organised search for value*. Longman, London.

Dell'Isola, A., (1997) *Value Engineering: practical applications for design, construction, maintenance and operations*. R. S. Means Co, Kingston, Mass.

Duerk, D. P. (1993) *Architectural Programming: Information Management for Design*. Wiley, Chichester.

Egan, Sir John (1998) *The Egan Report: Rethinking Construction*. Department of the Environment, Transport and the Regions, London, Publications Sale Centre, Rotherham, England.

Fallon, C. (1980) *Value Analysis*, 2nd edn. Lawrence D. Miles Value Foundation, Washington D. C.

Fellows, R., Langford, D., Newcombe, R. & Urry, S. (2002) *Construction Management in Practice*, 2nd edn, Blackwell Publishing Ltd, Oxford.

Gage, W. L. (1967) *Value analysis*. McGraw-Hill, London.

Glaser, B. G. & Strauss, A. L. (1967). *The Discovery of Grounded Theory: strategies for qualitative research*. Weidenfeld & Nicolson, London (printed in USA).

Green, S. D. (1992) *A SMART Methodology for Value Management*. Occasional Paper No. 53. Chartered Institute of Building, Ascot, Berkshire.

Hall, R. H. (1969) *Occupations and the Social Structure*. Prentice Hall, Harlow, Essex.

Hayes N. (2002) *Managing Teams: A Strategy For Success*, 2nd edn. Thomson Learning, Mitcham.

Hellriegel, D., Slocum, J. W. & Woodman, R. W. (1998) *Organizational Behaviour*, 8th edn. South-Western College, Cincinnati, Ohio.

Hines, P., Lamming, R., Jones, D., Cousins, P. & Rich, N. (2000) *Value Stream Management, Strategy and Excellence in the Supply Chain*. Financial Times Prentice Hall, Harlow, Essex.

Hunt, J. W. (1972) *The Restless Organisation*. Wiley, Chichester.

Hunt, J. W. (1981) *Managing People at Work: a manager's guide to behaviour in organizations*. Pan Business/Management, London.

Hunt, J. W. (1992) *Managing People at Work: a manager's guide to behaviour in organizations*, 3rd edn. McGraw-Hill, London.

Janis, I. L. (1972) *Victims Of Groupthink: A Psychological Study Of Foreign Policy Decisions And Fiascos*. Houghton Mifflin, Boston, Mass.

Janis, I. L. & Mann, L. (1977) *Decision-Making: A Psychological Analysis of Conflict, Choice and Commitment*. Free Press. New York.

Johnson, G. & Scholes, K. (1999) *Exploring Corporate Strategy: Text and Cases*, 5th edn. Prentice Hall Europe, Hemel Hempstead.

Juran, J. M. & Gryna, F. M. (1988) *Juran's Quality Control Handbook*, 4th edn. McGraw-Hill, New York.

Kaplan, R. S. & Norton, D. P. (1992) *The Balanced Scorecard*. Harvard Business Review January/February, Harvard Business School, Boston, Mass.

Kaufman, J. J. (1990) *Value Engineering for the Practitioner*. North Carolina State University, Chapel Hill, NC.

Kay, J. (1993) *Foundations of Corporate Success: How Business Strategies Add Value*. Oxford University Press, Oxford.

Kay, J. (2000) Strategy and the Delusion of Grand Designs in Mastering Strategy, pp. 5–16. *Financial Times*.

Kelly, G. (1955) *The Psychology of Personal Constructs*, Vols 1 & 2. Norton, New York.

Kelly, J. R. & Male, S. P. (1986) *A Study of Value Management and Quantity Surveying Practice*. Surveyors Publications, London.

Kelly, J. R. & Male, S. P. (1991) *The Practice of Value Management: Enhancing Value or Cutting Cost?* Royal Institution of Chartered Surveyors, London.

Kelly, J. & Male, S. (1993) *Value Management in Design and Construction: The Economic Management of Projects*. E. & F. N. Spon, London.

Kelly, J., Macpherson, S. & Male, S. (1992) *The Briefing Process: A Review and Critique*, Royal Institution of Chartered Surveyors, London.

Kelly, J., Male, S. & Macpherson, S. (1993) *Value Management: A Proposed Practice Manual for the Briefing Process*, Royal Institution of Chartered Surveyors, London.

Kirk, S. J. & Spreckelmeyer, K. F. (1993) *Enhancing Value in Design Decisions*. S. J. Kirk, Detroit, USA.

Langford, D. & Male, S. P. (2002) *Strategic Management in Construction*. Blackwell Publishing, Oxford.

Locke, K. (2001) *Grounded Theory in Management Research*. Sage Publications, London.

Macedo, M. C. Jr., Dobrow, P. V. & O'Rourke, J. J. (1978) *Value management for construction*. Wiley Interscience, Chichester.

Male, S. P. (1990) Professional authority, power and emerging forms of 'profession' in quantity surveying. *Construction Management and Economics*, **8**, No 2, pp. 191–204

Male, S. P. (1991) Strategic management and competitive advantage in construction. In: *Competitive Advantage in Construction* (eds S. P. Male & R. K. Stocks), pp. 45–104. Butterworth-Heinemann, Oxford.

Male, S. P. (2002) Supply chain management. In: *Engineering Project Management*, (ed N. J. Smith), 2nd edn. Blackwell Publishing Ltd, Oxford.

Male, S. & Kelly, J. (1989) Organisational responses of public sector clients in Canada to the implementation of value management: lessons for the UK construction industry. *Construction Management and Economics*, **7**, No. 3, 203–216.

Male, S., Kelly, J., Fernie S., Gronqvist M., Bowles G. (1998) *The Value Management Benchmark: A Good Practice Framework for Clients and Practitioners*. Published Report for the EPSRC IMI Contract. Thomas Telford, London.

Male, S., Kelly, J., Fernie, S., Gronqvist, M. & Bowles G. (1998) *The Value Management Benchmark: Research Results of an International Benchmarking Study*. Published Report for the EPSRC IMI Contract. Thomas Telford, London.

McGeorge, D. & Palmer, A. (1997) *Construction Management: New Directions*. Blackwell Science, Oxford.

McNiff, J. (2000). *Action Research in Organisations*. Routledge, London.

Miles, L. D. (1989) *Techniques of Value Analysis and Engineering*, 3rd edn. Lawrence D. Miles Value Foundation, Washington D. C.
Morgan M. H. (1960) *Vitruvius: The Ten Books on Architecture*, Dover Publications Inc., New York.
Morris, P. W. G. (1997) *The Management of Projects*. Thomas Telford, London.
Morris, P. W. G. & Hough, G. H. (1987) *The Anatomy of Major Projects: A Study of the Reality of Project Management*. Wiley, Chichester.
Mudge, A. E. (1989) *Value Engineering: A Systematic Approach*. J. Pohl Associates, Pittsburgh.
Norton, B. R. & McElligot, W. C. (1995) *Value Management in Construction: A Practical Guide*. Macmillan, Basingstoke.
O'Brien, J. J. (1976) *Value analysis in design and construction*. McGraw-Hill, London.
Parker, D. E. (1977) *Value Engineering Theory*. Lawrence D. Miles Value Foundation, Washington D. C.
Pena, W., Kelly, K. & Parshall, S. (1987) *Problem Seeking*, 3rd edn. AIA Press, Washington.
Pirsig, R. M. (1991) *Zen and the Art of Motorcycle Maintenance*. Vintage, London
Porter, M. E. (1985) *Competitive Advantage: Creating and Sustaining Superior Performance*. The Free Press, New York.
Reading Construction Forum (1996) *Value For Money*. Centre for Strategic Studies, Reading.
SAVE International (1988) *Value Methodology Standard*. Revised October 1998. Society of American Value Engineers, Dayton, Ohio, USA.
Schein, E. H. (1980) *Organisational Psychology*, 3rd edn. Prentice-Hall, Englewood Cliffs.
Shillito, M. L. & De Marle, D. J. (1992) *Value: Its Measurement, Design and Management*. Wiley, Chichester.
Snodgrass, T. J. & Kasi, M. (1986) *Function Analysis: the Stepping Stones to Good Value*. University of Wisconsin, Madison, Wisconsin.
Standing, N. (2001) *Value Management Incentive Programme: innovations in delivering value*. Thomas Telford, London.
Thiry, M. (1997) *Value Management Practice*. PMI Publications, Philadelphia.
Vorley, G. (1998) *Quality Management: Principles and Techniques*, 3rd edn. Quality Management and Training (Publications) Ltd, Guildford.
Winch, G. M. (2002) *Managing Construction Projects*. Blackwell Publishing, Oxford.
Woodhead, R. and Downs, C. (2001) *Value Management: improving capabilities*. Thomas Telford, London.
Woodhead, R., & McCuish, J. (2003) *Achieving Results*. Thomas Telford, London.
Zimmerman, L. W. & Hart, G. D. (1982) *Value Engineering: a practical approach for owners, designers and contractors*. Van Nostrand Reinhold, London.

Index

Accelerating Change Report, 2
ACID test, 53, 89, 274
action plan, 275
activities, client, 171
adjacency matrix, 72, 305
aesthetic, 12
attitudes, 148
audit, 25, 118, 275
Australia, 32, 222

balanced scorecard, 204, 264
Baldridge National Quality Award, 201
Belbin, 86
beliefs, 148
best value, 259
bid conference, 119
BOOT (build own operate transfer), 19
brainstorming, 56, 77
briefing, 118
BS EN 12973:2000, 33–6, 221, 228, 258
budget, 15
business value, 162

change management, 55, 98, 141, 278
charette, 18, 20, 85, 113, 129
checklist, 276
client business, 156
client definition, 1
client needs, 15
client value chain, 178
client value system, 56, 125, 157, 170, 171, 178, 206, 278–81
client wants, 15
community, 54, 277
concept design study, 111

concurrent study, 27
Construction Best Practice Panel (CBPP), 206
Construction Industry Board, 2
contractor's change proposal, 26
corporate value, 162
cost definition, 17
critical success factors, 127
customer features, 12

decision levels, 40
decision to build, 188
defence estates, 2
demand chain, 186
Deming, 199–201
design quality indicator, 205
design to cost, 281
document analysis, 281
drivers analysis, 136

education and training, 245–7
Egan Report, 2
element definition, 38, 73
element function analysis, 138, 282
element function list, 75, 283
elemental cost planning, 73
esteem value, 208
ethics, 247–50
European Foundation for Quality Management (EFQM), 202–3, 264
exchange value, 208

facilitation definition, 90
facilitation styles, 95–7, 133
facilitator, 78
facilities programming, 118

facilities walkthrough, 282
failure mode and effects analysis, 283
final sketch design study, 114
finance, 54
flowcharting, 71
forty hour study, 21–4
Function analysis system technique (FAST), 15, 60–71, 135, 283
function analysis, 51–80, 128, 136, 236, 257
function basic, 17
function definition, 17
function diagramming, 60–71
function element, 75
function secondary, 17
function space, 70, 138, 283
function strategic, 60, 67
function technical project, 66
functional performance specification, 283

gates hard, 141
gates soft, 141
gateway process, 2, 33
goal and systems modelling, 286
group decision support, 117
group definition, 81
group permanent, 82
group temporary, 82

idea reduction, 287
impact mapping, 288
implementation phase, 103, 139, 258
Institute of Value Management, 11, 236
international benchmarking study, 36–48
interviews, 124, 289
issues analysis, 55, 127, 135, 136, 290

Japanese VE (value engineers), 31
job plan, 13, 103, 122
Juran, 195

Kano, 197–8, 291
Kaufman, 66–70

Latham Report, 2
leadership, 87, 134

lean construction, 181, 184
legal issues, 55, 277
lever of value, 292
life cycle costing, 293
life of component, 12

Miles, Lawrence, 11
Movement for Innovation (M4I), 2

New York City Office of Management and Budget, 14, 26
NHS, 2

occupations classification, 238–9
operations workshop, 116
options appraisal, 295
organisation client, 54, 276
organisational change studies, 117
orientation and diagnostics, 26, 52, 103, 123, 258
outline sketch design study, 111

paired comparison, 209
paradigms and perspectives, 180
partnering, 119, 190
performance indicators, 205
politics, 54, 277
post occupancy evaluation, 293
PPC 2000, 2
presentation, 136, 294
Prime Contract, 2
Private Finance Initiative (PFI), 19, 169, 180, 188, 190, 212
ProCure 21, 2
procurement, 19, 118, 157, 187, 190, 212
professionalism, 239–44
programmes of projects, 59, 117, 159, 175, 177
project administered, 171
project briefing study, 109
project completion definition, 57
project constraints, 15, 55, 278
project context, 54, 277
project definition, 51, 170
project drivers, 55, 57, 281, 294

project execution plan, 33, 295
project finance, 277
project location, 54, 277
project teams, 83
project value chain, 178, 188
project value system, 169, 170, 171

quality, 12, 195
quality assurance, 199, 203
quality continuum, 197
quality definition, 195, 199
quality experience, 195
quality function deployment, 296
quality knowledge, 195
quality, lever of, 28
quality, TQM, 199

REDReSS (reorganisation, expansion, disposal, refurbishment, safety and security), 58, 299
rethinking construction, 2
risk, 138, 299–302

SAVE International, 13, 21, 222, 236, 252, 258
site tour, 302
Six Sigma, 204, 264
SMART (simple multi-attribute rating technique), 137, 302–5
Society of American Value Engineers, 12
space specification, 71, 73
stakeholder, 276
strategic briefing study, 107, 120
strategy choice, 155
strategy formulation, 155, 172
strategy implementation, 155
superior-subordinate issues, 56
supply chain management, 186
SWOT (strengths, weaknesses, opportunities and threats), 138, 307

TAM 2000, 32
team definition, 83

team dynamics, 84, 131
team independent, 143
team of record, 143
team size, 84, 133
time, 15, 54, 277
time, cost, quality triangle, 207, 307
timeline, 57, 308
Total Quality Management (TQM), 199
traditional procurement, 190, 212
turnkey procurement, 190

US Department of Defence, 12
US Government patronage, 13
use, 12
use value, 208
users, 69, 308

value analysis definition, 12, 237, 308
value chain client, 178
value corporate / business, 175
value customer, 186
value definition, 1, 17, 147
value engineering definition, 165
value engineering study types, 18, 21–7
Value for Europe Certification, 221
value management definition, 1, 29–31, 123, 165, 228
value management practice, 235
value operational, 175
value opportunity points, 235
value systems, 126, 128, 130, 141, 153, 178, 209
value thread, 130, 141, 159, 162, 169, 177
Vitruvius, 210

weighting and scoring, 308–10
whole life costing, 311–14
working groups, 132
workshop agenda, 125, 127, 131
workshop duration, 130
workshop management, 93
workshop phase, 103, 258
worth definition, 17